High Altitude Sickness – Solutions from Genomics, Proteomics and Antioxidant Interventions

D1809752

Narendra Kumar Sharma • Aditya Arya
Editors

High Altitude Sickness – Solutions from Genomics, Proteomics and Antioxidant Interventions

 Springer

Editors
Narendra Kumar Sharma
Department of Bioscience and
Biotechnology
Banasthali Vidyapith
Tonk, Rajasthan, India

Aditya Arya
National Institute of Malaria Research, Indian
Council of Medical Research
New Delhi, India

ISBN 978-981-19-1010-4   ISBN 978-981-19-1008-1   (eBook)
https://doi.org/10.1007/978-981-19-1008-1

This Springer imprint is published by the registered company Springer Nature Singapore Pte Ltd.
The registered company address is: 152 Beach Road, #21-01/04 Gateway East, Singapore 189721,
Singapore

Preface

The mighty Himalayas and several other high mountains across the globe represent marvellous creations of nature which have always attracted the attention of humans by their mighty heights and breathtaking views. They have always been an abode of some of the most scenic and adventurous trekking and tourist destinations, some are, however, places of religious importance, and others form strategic borders across nations. Some of the international borders lying within these mighty mountains, such as India-Pakistan and Indo-China borders, are highly strategic and have routine deputation of soldiers and army personnel. On the one hand, while the beauty of these mighty elevations attracts people, tough terrain and reduced density of air due to their altitudes pose challenges for high-altitude travellers. Some of the human establishments at the astonishing heights of above 5500 m are known to cause drastic physiological changes which sometimes culminate in lethality. High-altitude pulmonary oedema (HAPE), high-altitude cerebral oedema, loss of cognitive functions, hypertension and other hemodynamic abnormalities are key problems that need a clinical resolution. Moreover, identification of susceptibility and adaptability contributing factors are warranted to be resolved. Researchers across the globe over several decades have therefore explored the molecular, and physiological bases of these pathologies and undermined various mechanisms, effects, and outcomes. Proteomics and genomics remain a highly promising domain to explore the aforementioned problems, and considerable success has been achieved in terms of biomarker discovery and deciphering the precise sequence of molecular events occurring during the hypobaric hypoxia exposure. This book is aimed at providing a current perspective of genomic and proteomic approaches implied to hypoxia research and also an exploration of various possible interventions for minimizing the physiological alterations. This book includes views of experts across the globe working on simulated hypoxia models or human subjects travelling to high altitudes and brings along their decades of experience. We believe that the text would be useful for novice to experienced researchers and aid them in bringing refined hypotheses for hypoxia-induced pathophysiological changes and shall be the lead reference for the development of clinical solutions in the coming time. We acknowledge the defence research and development organisation, New Delhi, for valuable inputs and authors' contributions which enabled us to bring the book to its present shape. Any suggestions, comments, or critiques are welcome for the improvement of the book.

Tonk, Rajasthan, India   Narendra Kumar Sharma
New Delhi, India   Aditya Arya
December 2021

About the Book

This book illustrates the importance and significance of the proteomics studies in the domain of high-altitude physiology and associated sickness. More than 140 million people reside in high-altitude regions around the globe. A significant number of low landers visit high altitudes for professional activities, adventure sports, and leisure compounding occurrence of high-altitude illnesses. Hence, this book is intended for a global audience, health authorities, researchers, as well as professional sportsperson and coaches, to cover disorders and sickness associated with short-term and long-term high-altitude stay. The book starts with a brief introduction of high-altitude pathophysiology its cause and symptoms and later followed by a deeper understanding of ongoing research to develop proteomic based biomarkers for non-invasive diagnosis of altitude sickness susceptibility and also proven prophylactic and anaphylactic interventions. Therefore, this book will provide the reader a quick insight into contemporary methods in high-altitude proteomics and redox biology and new potential diagnostic and therapeutic strategies for clinicians for carrying forward to clinical trials. The book contains ten chapters, which include a sequential ascent of the text. The first few chapters have a primary focus on understanding the problem, while the latter half of the books includes discussion on various proteomic and molecular solutions emerging from current research. Self-explanatory illustrations and graphics, as well as research data from respective authors, have been used to enhance comprehension.

Contents

About the Editors

Narendra Kumar Sharma is working as Assistant Professor at the Department of Biosciences and Biotechnology of the Banasthali Vidyapith, India. He has earlier served as Assistant Professor in the Science Department at AKM, Kota University, India, Postdoctoral Fellow at Guangzhou Institutes of Biomedicine and Health (GIBH), Chinese Academy of Sciences (CAS), China, and Postdoctoral Fellow at Federal University of Sao Paulo (UNIFESP), Brazil. His research interest is in Genomics and Proteomics with specialization in hypoxia physiology, oxidative stress, posttranslational modification, protein biochemistry, signalling pathway, protein–protein interaction, infectious disease, regenerative medicine, and genome editing. He has been conferred with the prestigious award Dr. T. S. Vasundhara memorial best paper award, DRDO, India. He has served as a referee for a number of international journals. He has also published more than 20 research articles in peer-reviewed international journals and authored or co-authored books and book chapters. He has presented his research work at various international conferences held in India, Brazil, China, USA, and Germany. He is a member of many international scientific societies and organizations importantly, Indian Academy of Neuroscience (IAN), India, and International Organization of Scientific Research (IOSR), India.

Aditya Arya is working as a scientist at the National Institute of Malaria Research, New Delhi, a flagship lab of the Indian Council of Malaria Research (ICMR). Previously, he worked as a staff scientist at Pathfinder Research and Training Foundation, Gautam Budh Nagar, India. His research interest is in nanomedicine and redox biology. He has been trained at some of the most prestigious institutions such as Wellcome Trust Sanger Institute, IBRO, EMBL and holds active membership in several research societies. He has published more than 20 research papers in peer-reviewed journals about redox biology, proteomics, and high-altitude physiology. Besides this, he is also an academician and authored several book and book chapters like Concise Biochemistry, Understanding Enzymes, to name a few.

Introduction to High Altitude and Hypoxia

1

Preeti Sharma, Poornima Pandey, Pooja Kumari, and Narendra Kumar Sharma

Abstract

Climbing to a high altitude causes breathing to shorten, which reduces the amount of oxygen in the tissues and causes hypoxia. High altitude sickness is a medical illness with lethal implications such as hypoxia, high altitude pulmonary oedema (HAPE), high altitude cerebral oedema (HACE), and several other neurological disorders. Acclimatization is a primary response that encounters hypoxia and during this time individuals adapt to the decreased level of oxygen at a specific height. To investigate this physiological process, several studies have been conducted in the last few years. These studies have indicated the changes in the transcriptional and translational levels of various stress-associated genes/proteins under hypoxia and hypoxia acclimatization. Reducing air pressure at high altitudes causes hypoxia, which is a potential threat to the normal functioning of the brain. The generation of excessive free radicals and their intracellular diffusion leads to oxidative stress. Recent studies on molecular signalling along with shreds of evidence from cognitive impairment in the animal model during hypoxia have demonstrated that the cortex and hippocampus as anatomically and biochemically most vulnerable to oxidative stress in contrast to other regions of the brain. The emerging tools such as omics can be a milestone to study the physiological response of high altitudes and can decrease the adaptation time at high altitudes.

P. Sharma · P. Pandey · P. Kumari · N. K. Sharma (✉)
Department of Bioscience and Biotechnology, Banasthali Vidyapith, Tonk, Rajasthan, India
e-mail: snarendrakumar@banasthali.ac.in

Keywords

Hypoxia · High altitude pulmonary oedema · High altitude cerebral oedema · AMS

1.1   Introduction

Altitude is defined as vertical height above sea level and is directly associated with the relative air pressure present in the atmosphere. As the height increases, the air pressure decreases simultaneously and as the name indicates high means 'excessive' and altitude means 'elevation'. Thus high altitude (HA) is the term used for the specific vertical distance at which it becomes more difficult to survive and even may lead to life-threatening events because of the lack of availability of oxygen and some basic human body needs. Altitude is classified based on height above the sea level and categorized as high, very high, and extreme altitude. The specified distance of high altitude is about (8000–12,000 ft), very high (12,000–18,000 ft), and extreme (>18,000 ft, Fig. 1.1, Wilson et al. 2009).

People like to go to the mountains for various activities such as trekking and scenic views without knowing the fact that reduced oxygen at high altitude may

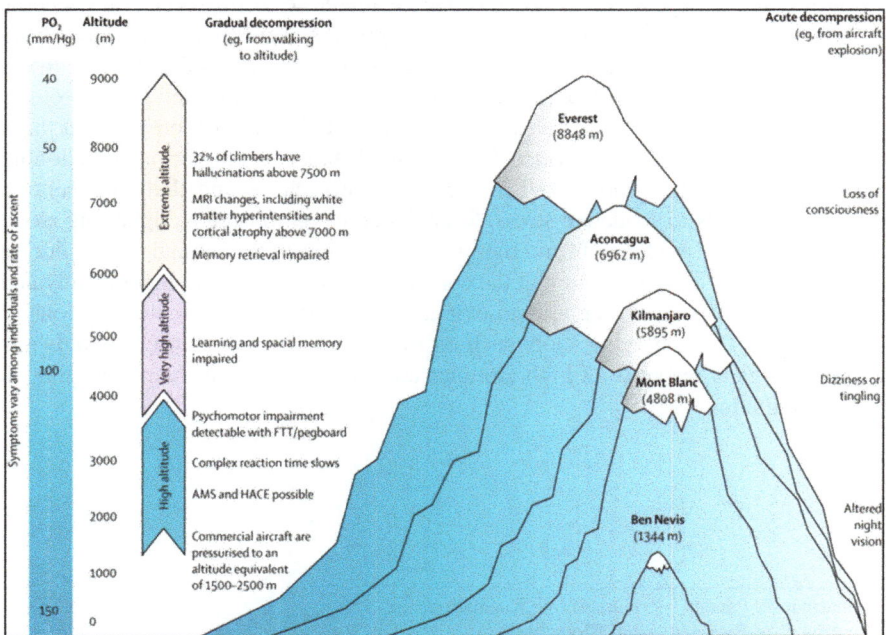

Fig. 1.1 The relationship between altitude and oxygen partial pressure, as well as the neurological effects of abrupt and progressive pressure shifts. (Adapted from Wilson et al. 2009)

affect them adversely. Hence, it is important to know the physiological consequences of HA to avoid HA associated disorders and life-threatening conditions. The aerial route nowadays becomes the prime choice of the middle-class family as it cuts the time of travel as well as easy availability of flights with affordable prices. These people reach the various touristic spots at high altitudes without acclimatization and start exploring HA. In the lack of knowledge, they develop HA sickness and other HA associated diseases. Even in this circumstance, aircraft would not operate over the Himalayan regions because of lower aerial pressure compared to required. The HA travellers have developed the acute mountain sickness (AMS) at the altitude of 2000 m and some of them have developed the HAPE and HACE when continued travel to HA (Dünnwald et al. 2021). Psychomotor retardation has occurred at very high altitudes that affected walking movement, talking, and other basic activities. Furthermore, impairment of learning and special memory has resulted in confusion or disorientation. At extreme altitudes, travellers fail to recall memory. The travellers have felt illusions at above 7500 m and other symptoms such as night vision alteration, dizziness or tingling and loss of consciousness.

The mortality rate is low, i.e. 4% at high altitudes but it is still important as soldiers reside most of the time at high altitudes. Each year, several pilgrims have died due to HA during the holy pilgrimage. There were more than 130 deaths reported yearly. The highest mortality was found at the death zone of Mount Everest and it increases with the increase in the number of climbers proportionally. When we talk about the HA, India has a long, sensitive border along with the world's highest mountain range (Khanna et al. 2018). The world's highest battlefield also lies in India, Pakistan, and China. The soldiers have to stay from 9000 to 20,000 ft for occupying their positions and it affects both mental as well as a physical condition due to the development of hypobaric hypoxia. The term hypobaric hypoxia (HH) is used when oxygen is limited due to reduced pressure at altitude. HH is a major physiological threat during the stay at a high altitude. It is associated with the increased number of RBSs and hemoglobulin to counter the low oxygen. There are no specific factors such as age, sex, and capacity that correlate with altitude sickness.

1.2   High Altitude and Oxygen Availability

The air is present up to 10,000 m at the end of the troposphere. The air is a mixture of gases in a definite concentration, for example, oxygen is present around 21% in the air. Most people are confused with the percentage of oxygen availability at HA. Atmospheric pressure is low at high altitudes when compared to sea level due to the gravity that pulls the air as close as to the ground. In case of high altitude (HA) the oxygen percentage remains constant, while the partial pressure changes according to elevation. So the partial pressure of oxygen lowers as the barometric pressure decreases, that cause the reduction of oxygen at tissue level. The gaseous exchange depends on the difference of the partial pressure. Oxygen exchange is occurred in lung cells and is carried by haemoglobin. One gram of haemoglobin carries 1.39 mL of oxygen. The level of haemoglobin increases to enhance the oxygen availability for the adaptation at HA. Once the oxygen availability is limited at HA, the heart rate is affected by blood viscosity, resulting in CMS (chronic mountain sickness) in which the blood pressure rises as the altitude rises (Crocker et al. 2020). Most of the travellers face AMS that is very common at HA with mild symptoms around 10,000 ft. The extreme altitude around 18,000 ft or higher cannot be tolerable without oxygen supplementation. With the elevation of height, physiological changes occur in our bodies. The first system that responds to HA is the respiratory system. The atmospheric pressure at sea level is 760 mmHg that corresponds to approximately 159 mm of partial pressure of oxygen (PO_2). During respiration, inhaled air is warmed and humidified that resulted in the addition of water vapour in the inhaled air and affects the partial pressure of the gases. At sea level, the oxygen levels in the blood within the capillaries exceeds that found inside the alveoli about 0.25 s of inhaling air (Zubieta-Calleja and Zubieta-DeUrioste 2021). Furthermore, while PAO_2 levels are lower at higher elevations, the pace of building oxygen tension in capillary blood is slower than at sea level. At rest, the capillary transit time is still sufficient to allow the oxygen tension to approach that within the alveoli, but when the transit time is reduced by exercise there will be a marked worsening of any hypobaric hypoxia. When PO_2 is decreased, chemoreceptors from the central nervous system (CNS) are stimulated in the dorsal respiratory groove that resulted in hyperventilation. It suggests that during hyperventilation, more oxygen is delivered to the HA for the acclimation process. Blood remains in the lungs for about 0.75 s. The atmospheric carbon dioxide level is also very less at high altitudes (Storz and Cheviron 2021).

1.3   Acclimatization at High Altitude

Acclimatization refers to the process where our body compensates for the low partial pressure of oxygen by either amplifying certain receptors or increasing the number of RBCs. A finger pulse oximeter is used to measure physiological consequences at HA as an acclimatization process. There are several physiological changes occurred in respiration and circulation when going to HA. By acclimatization process, we can reduce the deleterious effect of low oxygen on the body increases ventilation and circulation (Fig. 1.2).

When people travel to HA above 3000 m for various activities with a lack of knowledge of HA physiology, they have noticed several HA associated sicknesses including headache, dizziness, nausea, hyperventilation, etc. When they did not acclimate at HA and have begun their activities there, their symptoms may intensify, and in rare circumstances, life-threatening scenarios will occur. Some of the travellers and soldiers have developed pulmonary oedema/cerebral oedema. In this case, they have to come back at sea level immediately, however, now we have some medications so the treatment can be done at HA itself. Rapid ascent to HA is associated with decreased health and fitness, while they feel normal after returning to sea level. The acclimatization process helps the travellers to adopt the adverse

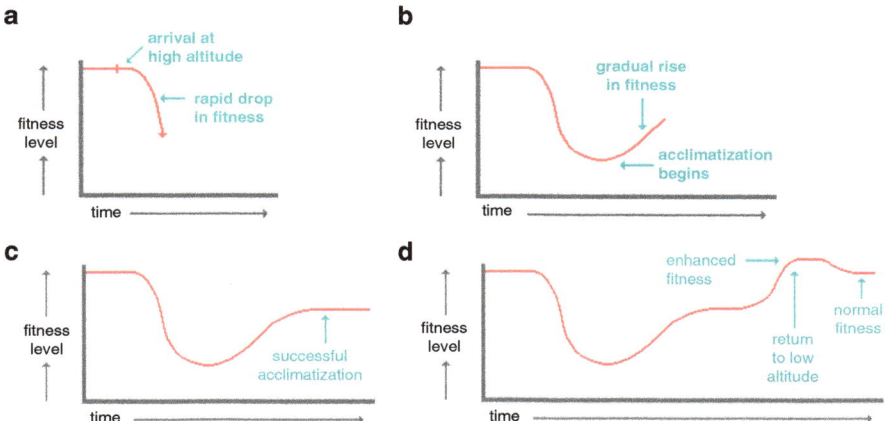

Fig. 1.2 Process of acclimatization. (**a**) Initial inefficient response to low oxygen pressure (**b**) Beginning of successful acclimatization to low oxygen pressure (**c**) Increased fitness level after successful acclimatization to low oxygen pressure and (**d**) Enhanced fitness level for a short period after returning to low altitude. (Source: http://anthro.palomar.edu/adapt/adapt_3.htm)

conditions quickly at HA and reduces the risk of development of AMS, HACE, and HAPE (Bärtsch and Swenson 2013). The successful acclimatization process has been reported at 5500 m, further ascent after this height depends on the individual's physiology and immunity (Fig. 1.3).

Slow stepping and staging is the most common strategic event for high altitude sickness and speedy acclimatization, as the cliché goes, slow and steady wins the race. After ascending to 1000 m, it is typically recommended that HA travelers do not venture out 500 m per day. During ascension, staging refers to stopping for a period of time and at a specific altitude. Both of these processes of acclimatization are used in mountaineering as well as in trekking at HA. There are several protocols used for acclimatization like some exercise after resting at HA. Pre-acclimatization and intermittent hypoxia exposure strategies have worked well for high altitude sickness (Luks et al. 2017). Recently, members of the union international des association's d'Alpinisme medical commission (UIIA Med Com) claimed that high altitude climbers didn't found to be symptomatic of AMS as compared to the other ones. Pre-acclimatization strategies are also included in the carrying of drugs that minimize the effect of high altitude sicknesses such as Acetazolamide and Dexamethasone. These drugs along with some natural extracts such as a nasal cannula, *Biloba* also helped to decrease the deleterious effect of HA (Karinen 2013).

1.3.1 Acute Mountain Sickness

Acute mountain sickness is often known as "Mountain sickness" and "Altitude sickness," which is a clinical condition that can occur if people move or climb to a

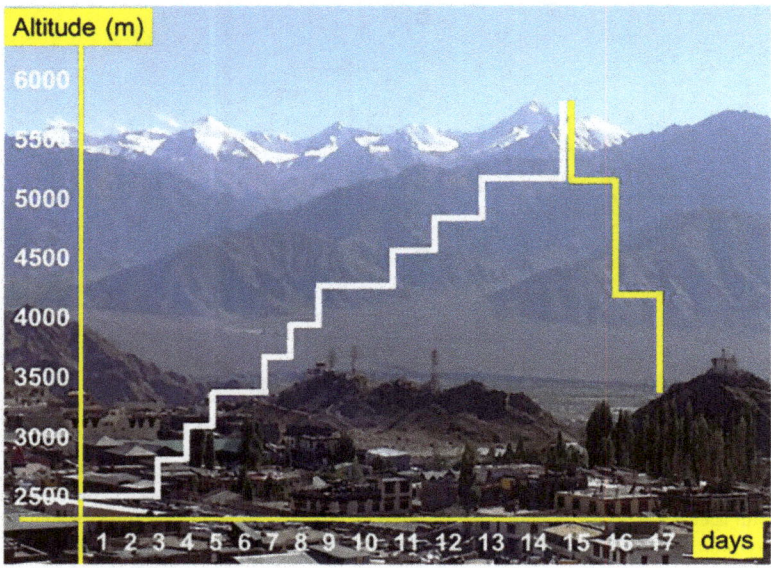

Fig. 1.3 Example of a safe (traditional) ascent from 2500 m to 6121 m (Stok Kangri, Ladakh). 300–500 m per day and one rest day for every 1000 m gain in altitude (Burtscher et al. 2021)

greater height or altitude too rapidly. This condition rise because of less oxygen availability like if someone went to a greater height the pressure dropped and oxygen availability dropped. This sickness can begin at any elevation above 8000 ft. AMS is characterized by headaches, dizziness, shortness of breath, drowsiness, poor appetite (often associated with nausea), exhaustion, fatigue, and irritability these symptoms may develop within 12–24 h after reaching high elevation and disappear within 2 days as the body adjusts to change in higher altitude (Wilson et al. 2009). In the case of mild altitude sickness symptoms may be more severe and will not improve with time and medicine, as time passes by instead of feeling good, a person will begin to feel bad and will have increased tiredness and breathlessness. HAPE (high altitude pulmonary oedema) and HACE (high altitude cerebral oedema) both are considered as a severe form of altitude sickness, induration, patients may acquire symptoms such as shortness of breath even when resting, inability to move, while coughing a white or pink substance foam, coma, and so on. Altitude sickness may affect anyone, no matter how strong, young, or physically fit they are, it can even affect athletes too. Various factors involved that can influence the risk of altitude illness include how rapidly people climb to a greater height, how high they go up and the altitude at which they sleep, it also depends upon where a person belongs (where to live) age and if they ever had altitude illness before. Some health issues also make it more severe to the people who suffer from these such as diabetes and lung disease, genes also play a role in the ability of the body to handle HA. The symptoms regarding HA sickness will assist in seeking treatment as soon as possible. The basic treatment for any altitude sickness is to descend as quickly as possible with staying safe, in case of severe altitude sickness person must be transferred to a lower height immediately. It should be less than 4000 ft, in the case of HACE steroid prescribe for the patient and in HAPE, the individual should be given oxygen and lowered as soon as possible. The other way to reduce high altitude illness is acclimatization; it might help to reduce the chances of experiencing altitude sickness as a person climb to HA they should gradually acclimatize their body to the variations in atmospheric pressure. Another way is to wear loose clothing to allow easy release of gastrointestinal gases similarly; some other recommendation is given to person going for high elevation like less utilization of milk and milk derivatives either switch to lactose-free diet on first of arrival good oral hydration is necessary at this time (Zubieta-Calleja and Zubieta-DeUrioste 2021). HAPE and in extreme instances HACE, both followed by AMS, can arise on the day of arrival and around 3 to 4 days later. The identification of mountain sickness is clinical because the patient is new to a higher altitude or not.

1.3.2 High Altitude Cerebral Oedema (HACE)

HACE (high altitude cerebral oedema) is a life-threatening neurological disease that develops in people with AMS or HAPE over hours to days. It is characterized by variable levels of disorientation, behavioural problems, paralysis, and cerebral oedema on neuroimaging. If it is not treated promptly, it can quickly progress to stupor, which can be deadly. Unlike HACE, AMS is characterized by vague symptoms and these two are generally linked with non-acclimatized persons who abruptly climb over 2500 m as a result, it is considered as the last stage of AMS (Hultgren 1997; Yarnell et al. 2000; Hackett 1999a, b). HACE could emerge 3 to 5 days upon arriving at elevations as low as 2700 m (9020 ft), although it is more commonly observed in isolated places higher than this range when the symptoms appear much more rapidly within an hour. Usually, it affects individuals of different ages and genders; however, youngsters may be more vulnerable due to continued rise despite AMS symptoms and a faster pace of ascension that lead to caused hypoxia (Jacob et al. 2020). HACE is diagnosed clinically, but the symptoms associated from mild to moderate. AMS can appear soon after the rise and typically worsen over 24 to 72 h. Notably, the peak period of AMS appearance often corresponds with the onset of HACE symptoms indicate a possible continuity from AMS to HACE (Sutton 1992).

The diagnosis of cerebral oedema is routinely done by neuroimaging investigations. The intracranial pressure, cerebrovascular, and MRI investigations have provided a better knowledge of the macroscopic changes associated with cerebral oedema. However, the forms of oedema may differ in different parts of the brain, in different individuals and it is not always evident in imaging through the processes. The mechanism of these changes at the vascular and cellular levels remain unclear (Jacob et al. 2020). The aetiology behind cerebral oedema includes an excessive build up of water in the parenchyma of the brain including various compartments such as interstitial and intracellular. Some of the evidence recommended that this type of oedema occurs when AMS develops. AMS and HACE both share a common pathological process include increasing cerebral blood volume and early formation of intracellular oedema, which is dependent on the osmotic gradient and further take part in the formation of ionic oedema alongside vasogenic oedema. Furthermore, several types of physiological processes have been recognized that take part in altered brain water regulation (Turner et al. 2021).

1.3.2.1 Intracellular Oedema

Intracellular oedema is osmotically dependent on the cellular movement of water and ions from cerebral extracellular to intracellular space. It is occurred due to the failure of an energy-dependent mechanism of transport, as a result, the intracellular space swell but there is no initial brain swelling was observed. It is shown that no additional volume was introduced into the intracranial section. This condition is also common in the periventricular white matter of those who have just moderate hypoxia. It is shown that energy-dependent pathways become impaired even at a moderate level of hypoxia (Turner et al. 2021).

1.3.2.2 Ionic Oedema

Intracellular oedema is reduced Na^+ Cl^- and water in the cerebral extracellular space and create an osmotic pressure gradient for this molecule across the capillary blood-brain barrier. As Na^+Cl^- homeostasis is restored in the cerebral extracellular space, water is pulled into the brain and developed extracellular /ionic oedema and this kind of oedema develop without changes or disruption of the blood–brain barrier (BBB) (Turner et al. 2021).

1.3.2.3 Vasogenic Oedema

In vasogenic oedema, cerebral extracellular fluid accumulation has occurred. It contains plasma proteins, that are largely linked to dysfunction of endothelial, vascular damage, and disrupt the integrity of the blood–brain barrier (BBB). It is found that the blood–brain barrier lost integrity in several cases of HACE (Turner et al. 2021).

Several processes are involved to understand the mechanism of action during HACE; it is started with cell progression in which the transition from artery to a vein occurred (left to right) and this progression is further increased intravascular pressure caused vasogenic oedema and arterial wall damage. Specific alterations have been found in the artery and vein, with increased hydrostatic pressure in the artery and venous outflow blockage in the vein. These partial pressures of oxygen and carbon dioxide have been thought to have direct vasoactive properties with hypoxemia producing vasodilation and hypocarbia causing vasoconstriction. The hypoxic ventilatory response has mediated a balance between hypoxemia and hypocarbia. Direct hypoxia may further induce Na^+/K^+ ATPase failure that caused cytotoxic oedema. Besides various chemical mediators that have been linked to the disease progression, free radical production is one of them that could cause vasogenic oedema by directly damaging vessel basement membranes. In addition to HIF-1 accumulation and consequent VEGF, overexpression could lead to more damage of the basal membrane and cerebral oedema. The limited hyperkalemia may also produce nitric oxide that is calcium-dependent and may further be caused vasodilation by acting on vascular smooth muscle. The vasodilation may also be caused by neuronally mediated adenosine release and activation of vasodilation is linked to the trigeminovascular system, which is caused the headache. In the complete scenario, micro-haemorrhage development is a key component of HACE, which can be generated by damage of vessels caused by cytokines or various other chemical mediators or by elevation of hydrostatic pressure (Wilson et al. 2009).

1.3.3   High Altitude Pulmonary Oedema (HAPE)

High altitude pulmonary oedema is a fatal disease; defined by the accumulation of fluid in the lungs as a result of acute high altitude hypoxic exposure. These clinical categories are also included AMS, which is more frequent and HACE. HACE is uncommon at HA and the complication of high altitude illness is occurred after above 3000 m. There are two main causes for the development of HAPE. First is pulmonary artery vasoconstriction which caused an increase in pulmonary circulation pressure and the second is an increase in capillary permeability that lead fluid to flow into the alveoli in patients who are prone to HAPE. It is generally followed by symptoms of AMS though the most common symptoms associated that are included shortness of breath, cough, dizziness, dyspnea, fatigue, and cyanosis.

Early symptoms of HAPE are included mild nonproductive cough, dyspnea with effort and reduced physical performance. These symptoms are often ignored by patient's that resulted in increased cough and severe dyspnea even at rest conditions. These extreme cases are distinguished by bubbling in the chest and pink foam of sputum. Although the exact prevalence is still not fully explained (Paralikar 2012), it is responsible for the majority of fatalities associated with high altitude disease and cases have been reported as low as 8500 ft approximately (3000 m) (Yarnell et al. 2000). The lack of early detection methods for HAPE also contributed to the complications at HA. This pulmonary oedema at high altitude is developed within a day to 3 days of arriving at HA, and then hardly after a day (Hackett and Roach 2001), it turns into HAPE. For this reason, hypoxia-induced extensive pulmonary vasoconstriction, capillary leak, and alternative diagnosis should be investigated. The imaging scans are also recommended by the hospitals after 4 days of arrival to HA. Various oxygen-sensing mechanisms such as nitric oxide syntheses, (VEGR) vascular endothelial growth factor, and hypoxia-inducible factors (HIF) are responsible for the activation of inflammatory, immunological, and physiological changes that cause HAPE.

HIF is a key transcriptional factor that stimulated the expression of various target genes that are responsible for the regulation of oxygen homeostasis (Woods and Alcock 2021). In recent years, the researchers are focused to study the signalling and biochemical processes in HAPE progression by emphasizing the fact that increased pulmonary hypertension on elevation leads to the development of HAPE. One hypothesis has claimed that minimizing pulmonary hypertension would prevent HAPE (Paralikar 2012). It is found that glucocorticoids have been proven to be useful in reducing HAPE in a person when administered before climbing and during ascent to HA. It is activated the production of cGMP in hypoxia and increased the activity of nitric oxide syntheses by activation of epithelial $Na^+ K^+$ ATPase pump, however, extensive research is still needed to find the lowest effective dosage. The treatment for HAPE is included quick enhancement of oxygenation by supplementary oxygen, hyperbaric treatment, or fast decline.

1.4   Hypobaric Hypoxia and Brain

Hypobaric hypoxia (HH) is a phenomenon in which organisms lack an adequate supply of oxygen to various body tissue at HA. Due to a decrease in partial pressure at HA, the ability to transfer oxygen from the lungs to the bloodstream is adversely affected. The effects of HH are associated with the height of an altitude. The effect of HH is indistinguishable in the early phase and the body response of the different individuals are also different. The most common symptoms are included as breathing difficulty, rapid pulse rate, lethargy, headaches, inability to think, and tiredness. Most of the body organs got affected by the limited availability of oxygen including the lung, heart, brain, renal, and spleen. Among all, the brain is the most vulnerable organ affected and it is the first organ to be impaired in the HH (West 1996; West et al. 2007).

The brain is the most studied organ under HH since it uses 20% of inspired oxygen and has substantial energy expenditures that must not be reduced. The brain cells consumed half of the energy for charge transfer across cellular membranes to maintain cellular redox equilibrium (Stieg et al. 1999). As a result of that, in a few minutes of oxygen deprivation, the brain experiences energy failure. Several studies have discussed the differences between high altitude (HA) natives and low lenders in terms of altitude-related brain alterations. According to Zhang et al., HA natives had less grey matter than sea level controls, which resulted in lower blood pressure (Zhang et al. 2010).

Because of the diminished oxygen flow to the brain, higher altitude hypoxia lowers a person's mental abilities (Lieberman et al. 2005), and the effect varies depending on altitude height. The lower ability for continuous mental effort, memory problems, audio and visual difficulties, and irritation are all possible neurological and cerebral symptoms that occurred at HA. Adults may experience weight loss of 5–10 pounds if their appetite remains low. The symptoms of the high altitude associated sickness may be persisted until the stay there that hinder the normal functioning of the body. In some circumstances, patients' need to return to sea level too, however with the proper acclimatization, these issues can be minimized.

Mitochondria are the major sources of reactive oxygen and nitrogen species. The superoxide is reacted with nitric oxide radicals and produced peroxy-nitrites that is considered strong oxidant. Under hypobaric hypoxia, the NO_2^- pathway is activated that produced excess free radicals through the respiratory chain. The immediate treatment for hypoxia is the supplementation of oxygen that can be given through an oxygen mask. HH has resulted in changes in blood flow to the brain, energy metabolism, and cognitive processes such as learning and memory. It is also responsible for cerebral damage but the molecular mechanism remains unknown yet.

Hypobaric hypoxia impairs cognitive function because of a variety of reasons including oxidative stress, neuronal damage, and neurotransmitter changes. Jayalakshmi et al. suggested that administration of N acetylcysteine improved the deleterious effects of hypoxia on spatial learning memory function. To diminish free radical production during hypobaric hypoxia, oxidative stress is reduced by increasing the antioxidant status (Jayalakshmi et al. 2007). Several other therapeutic agents are also reported to decrease cognitive impairment under hypobaric hypoxia such as acetyl-L-carnitine, L type calcium channel, and glutamate receptor (Barhwal et al. 2009), however, targeting single neural receptor, ion channels, and gene are unable to combat the enormous cascading of hypobaric hypoxia-induced alterations.

1.4.1 Brain as a Vulnerable Site of Oxidative Stress

The brain is extremely vulnerable to oxidative stress because of its high oxygen demand and lipid-rich composition. The intake of oxygen, on the other hand, led in the development of free radicals, sometimes known as the "essential devils of the cell." In a healthy organism, the balance of pro-oxidants and anti-oxidants is crucial and if it is shifted towards pro-oxidants, oxidative stress may occur. It is widely established that high altitude increases the generation of reactive oxygen species (ROS), which is linked to the pathophysiology of hypoxia-induced changes. Oxidative stress causes inflammation in the brain, which compromises the integrity of the blood-brain barrier, a key component of CNS homeostasis, and leads to brain injury. It also triggered neuronal death in the cortical, subcortical, and hippocampal regions of the brain (Sharma et al. 2011, 2013; Ahmad et al. 2013). Oxygen plays an important role in energy production in mitochondria, when acted as a limiting factor, excess formation of ROS is generated during cellular respiration (Gandhi and Abramov 2012). The generation of free radicals and ROS is determined by analyzing the mitochondrial resting or active staus. In a cell, ROS production is ten times higher in mitochondria than in cytosol and nucleus. Cytochrome oxidase complex (Complex IV) is responsible for oxygen sensing in normal conditions. During cellular respiration, complex IV transfers electrons to molecular oxygen and reduced them to water. ROS and free radicals are served as intermediate and are attached to the complex until completely utilized to form water. HH induced oxidative stress that leads to the generation of excess ROS and free radicals which degrade or oxidized macromolecules in the cell (Schild et al. 2003; Maiti et al. 2006; Halliwell 1989). The above-discussed events can be seen in Fig. 1.4.

Due to increased ROS production on the outer side of the inner mitochondrial membrane in HH, cytosol or IMS oxidation increased, whereas mitochondrial matrix oxidation reduced. Mitochondrial DNA (mt DNA) is more vulnerable to oxidative stress than nuclear DNA due to poor DNA repair function and the continuous

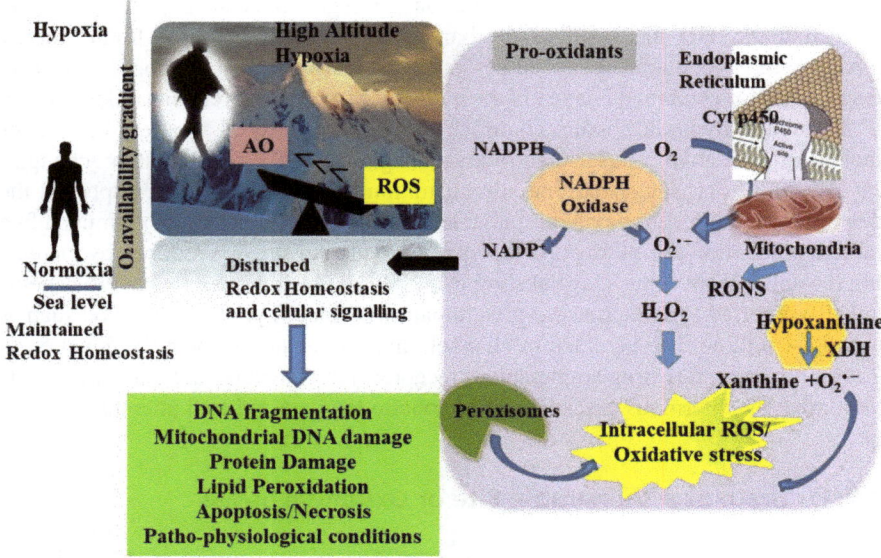

Fig. 1.4 Electron transport chain reaction showing oxygen as the terminal acceptor of electron and increases the level of ROS and the disturbance of redox homeostasis affects various biological reactions at the cellular level. (Adapted from Gaur et al. 2021)

production of free radicals which contributed to a highly reducing environment. Both the IMM and the IMS emit ROS during HH, which may lead to the activation of various transcription factors such as hypoxia-inducible factors (HIF1); additionally, FOXO-mediated transcription factors help to stabilize and influence a variety of biological responses in the body. Various studies revealed that a decreased level of GSH and elevated level of GSSG were also reported in humans to expose to high altitudes these are a good indicator of oxidative stress (Gaur et al. 2021). ROS are waste products of cellular oxidative metabolism (Halliwell 1989). They play critical roles as secondary messengers. However, increased ROS production overwhelms the antioxidant scavenging capability, resulting in oxidative damage to DNA, lipids, and proteins, as well as cellular damage (Bredesen 1995).

The brain also contains low to moderate levels of enzymes such as catalase, superoxide dismutase, and glutathione peroxidase, which are involved in the metabolism of reactive oxygen species (ROS) (Kankofer 2001; Işlekel et al. 1999; Dringen et al. 2000). The presence of iron in the brain and particularly in areas such as globus-pallidus and substantia nigra may also be contributed to the production of ROS.

1.4.2 Hypobaric Hypoxia and Memory Functions

Interactions between genes and the environment determine the behaviour. Learning and memory are the most significant mechanisms through which the environment influences human behaviour. Memory is the ability to retain and retrieve earlier sensations, ideas, experiences, or information that has been deliberately learnt, whereas learning is the process through which humans can gain knowledge about the universe. The cortex and hippocampus are the main regions that have a role in memory and learning. Any damage to these structures has an impact on cognition. HH is a severe disorder that resulted more frequently at high elevations and can damage cognitive performance. HH adversely affected the hippocampal neuron and resulted in the loss of spatial memory. Acute HH exposure at various altitudes resulted in a substantial loss in spatial memory, which worsened under chronic HH exposure. Several studies have been reported the various plant-based ingredients to prevent hypobaric hypoxia-induced spatial memory deficit. In addition to ROS, excess calcium caused mitochondrial uncoupling that reduced ATP production and subsequent neuronal death. Potassium (K+) channels have been discovered to be critical feedback for regulating the Ca_2^+ influx mechanism. Notably, apamin-sensitive Ca_2^+ controlled potassium channels triggered by high intracellular Ca2+ levels and contributed to hyperpolarization, cessation of neuronal activity, neuronal injury, and loss of memory functioning (Kushwah et al. 2021). Memory loss has been linked to neuronal death in the CA1 region of the hippocampus, whereas cortical neurons are equally prone to ischemia/hypoxia injury (Ruan et al. 2003). Neuronal death has occurred during HH exposure (Choi 1996). Necrotic signals such as ER growth, polyribosome disaggregation, and dendritic swelling have been reported in neurons in ischemia (Ruan et al. 2003). Hypoxia also leads to glial degeneration which indirectly affected neurons (Anderova et al. 2011). At HA (about 4500 m) short-term memory has begun to deteriorate and is mostly affected at 6000 m (Berry et al. 1989; Virués-Ortega et al. 2004), however, long-term memory seems to be preserved (Stobdan et al. 2008). The results of various studies are conducted on animals and humans indicated the fast impairment of the ability to learn under HH, rather than the ability to retrieve information (Kramer et al. 1993). HA around 3800 to 5000 m is also affected spatial memory. It is observed that the months after return from HA, cognitive flexibility was severely reduced in professional mountain climbers' using the Stroop colour and word test or the Wisconsin card-sorting test (Regard et al. 1989). Furthermore, different time durations at HH exposures have revealed a distinct pattern of histone acetyltransferases and histone deacetylases (HDACs) gene expression in the hippocampus. A study has revealed that the level of acetylation site in histone protein decreased during HH exposure. In addition, brain-derived neurotrophic factor (BDNF) has activated the cAMP-response element-binding protein (CREB) phosphorylation via the PI3K/GSK3/CREB axis, which mitigated hypobaric hypoxia-induced spatial memory impairment (Kumar et al. 2021).

1.5 Other Neurological Effects at High Altitude

Several other neurological disorders have occurred at HA that attained less importance including double vision, cerebral venous thrombosis, Ischemia strokes, fainting and convulsions, cortical blindness, and migraine. These disorders must be addressed well so as not misclassified with altitude sickness. The treatment of these disorders is also different from previously discussed (Basnyat et al. 2004). Elderly people are often suffered from cardiovascular disease under transient ischemic attacks at sea level. In contrast, the younger population is also suffered from ischemic attacks at HA without having any cardiovascular risk factors. The molecular mechanism is still not fully explored, however, some studies have shown that the alteration in vascular endothelial has resulted in vasospasm in the cerebral circulation. Some people have suffered from severe headaches, nausea, and dizziness during this condition.

Stroke is not observed at HA commonly but other factors made contributed significantly to it such as polycythaemia, cold, cough, and dehydration. In addition, coagulation irregularity is also added as a risk factor to stroke under HH (Wright et al. 2008). The HA travellers have felt the migraine. It is a highly prevalent phenomenon that affected both men and women at HA when compared to sea level. The symptoms are similar to AMS including severe headache at HA and oxygen-breathing therapy and CO_2 enriched air-breathing methods are used. The clinician has also prescribed drugs, i.e. aspirin in some cases for preventing migraine attacks at mountains. Presently, various advanced neuroimaging techniques are used to study the pathophysiology that will help to understand migraine at HA.

Another HA associated disorder is syncope that is commonly found in people who live at moderate altitudes. It is observed after ascent and appeared to be a vasovagal event due to alteration in blood pressure at HA. The symptoms are developed after the meal and/or drinking alcohol. However, it is not contagious and resumed to normal but precautions are needed when climbing to HA. In some cases, brain inflammation has occurred at HA. Studies have suggested that the increased intracranial pressure and swelling of the brain is contributed to the inflammation under HH exposure.

Few case studies have shown the sudden appearance of tumour cells with nausea and headache due to space-occupying lesions disorder (Yarnell et al. 2000). In other neurological alterations, blood coagulation is observed resulting from drugs intake to counter mountain sickness that further lead to thromboembolism. Some other ophthalmological problems are arisen such as retinal haemorrhage, cortical blindness, and amaurosis fauna. As the cerebral blood flow has suddenly arisen, it has increased endothelial permeability due to increased exertion intravascular pressure. Studies have shown the correlation of retinopathy with AMS, while the molecular mechanism is not fully known. In cortical blindness, pupillary reflex and visual cortex blood supply is compromised due to vascular spasm at HA. To overcome the situation, oxygen inhalation and in some cases rebreathing of carbon dioxide increased cerebral blood flow and improved this condition (Clarke 2006). The cornea receives oxygen directly from air by diffusion for normal functioning.

When eyes are closed, PO_2 falls and the central area gets flattened and caused uneven corneal swelling at HA.

References

Ahmad Y, Sharma NK, Garg I, Ahmad MF, Sharma M, Bhargava K (2013) An insight into the changes in human plasma proteome on adaptation to hypobaric hypoxia. PLoS One 8(7):e67548

Anderova M, Vorisek I, Pivonkova H, Benesova J, Vargova L, Cicanic M, Chvatal A, Sykova E (2011) Cell death/proliferation and alterations in glial morphology contribute to changes in diffusivity in the rat hippocampus after hypoxia-ischemia. J Cereb Blood Flow Metab 31(3): 894–907

Barhwal K, Hota SK, Jain V, Prasad D, Singh SB, Ilavazhagan G (2009) Acetyl-l-carnitine (ALCAR) prevents hypobaric hypoxia–induced spatial memory impairment through extracellular related kinase–mediated nuclear factor erythroid 2-related factor 2 phosphorylation. Neuroscience 161(2):501–514

Bärtsch P, Swenson ER (2013) Acute high-altitude illnesses. N Engl J Med 368(24):2294–2302

Basnyat B, Wu T, Gertsch JH (2004) Neurological conditions at altitude that fall outside the usual definition of altitude sickness. High Alt Med Biol 5(2):171–179

Berry DT, McConnell JW, Phillips BA, Carswell CM, Lamb DG, Prine BC (1989) Isocapnic hypoxemia and neuropsychological functioning. J Clin Exp Neuropsychol 11(2):241–251

Bredesen DE (1995) Neural apoptosis. Ann Neurol 38(6):839–851

Burtscher M, Hefti U, Hefti JP (2021) High-altitude illnesses: old stories and new insights into the pathophysiology, treatment and prevention. J Sport Health Sci 3(2):59–69

Choi DW (1996) Ischemia-induced neuronal apoptosis. Curr Opin Neurobiol 6(5):667–672

Clarke C (2006) Neurology at high altitude. Pract Neurol 6(4):230–237

Crocker ME, Hossen S, Goodman D, Simkovich SM, Kirby M, Thompson LM et al (2020) Effects of high altitude on respiratory rate and oxygen saturation reference values in healthy infants and children younger than 2 years in four countries: a cross-sectional study. Lancet Glob Health 8(3):e362–e373

Dringen R, Gutterer JM, Hirrlinger J (2000) Glutathione metabolism in brain: metabolic interaction between astrocytes and neurons in the defense against reactive oxygen species. Eur J Biochem 267(16):4912–4916

Dünnwald T, Kienast R, Niederseer D, Burtscher M (2021) The use of pulse oximetry in the assessment of acclimatization to high altitude. Sensors 21(4):1263

Gandhi S, Abramov AY (2012) Mechanism of oxidative stress in neurodegeneration. Oxid Med Cell Longev 2012:1–11

Gaur P, Prasad S, Kumar B, Sharma SK, Vats P (2021) High-altitude hypoxia induced reactive oxygen species generation, signaling, and mitigation approaches. Int J Biometeorol 65(4): 601–615

Hackett PH (1999a) High altitude cerebral edema and acute mountain sickness. A pathophysiology update. Exp Med Biol 474:23–45

Hackett PH (1999b) The cerebral aetiology of high-altitude cerebral edema and acute mountain sickness. Wilderness Environ Med 10(2):97–109

Hackett PH, Roach RC (2001) High-altitude illness. N Engl J Med 345(2):107–114

Halliwell B (1989) Free radicals, reactive oxygen species and human disease: a critical evaluation with special reference to atherosclerosis. Br J Exp Pathol 70(6):737

Hultgren HN (1997) High altitude medicine. Hultgren, Stanford, CA

Işlekel S, Işlekel H, Güner G, Özdamar N (1999) Alterations in superoxide dismutase, glutathione peroxidase and catalase activities in experimental cerebral ischemia-reperfusion. Res Exp Med 199(3):167–176

Jacob DW, Ott EP, Baker SE, Scruggs ZM, Ivie CL, Harper JL, Limberg JK (2020) Sex differences in integrated neurocardiovascular control of blood pressure following acute intermittent hypercapnic hypoxia. J Physiol Regul Integr Comp Physiol 319(6):R626–R636

Jayalakshmi K, Singh SB, Kalpana B, Sairam M, Muthuraju S, Ilavazhagan G (2007) N-acetyl cysteine supplementation prevents impairment of spatial working memory functions in rats following exposure to hypobaric hypoxia. Physiol Behav 92(4):643–650

Kankofer M (2001) Antioxidative defence mechanisms against reactive oxygen species in bovine retained and not-retained placenta: activity of glutathione peroxidase, glutathione transferase, catalase and superoxide dismutase. Placenta 22(5):466–472

Karinen H (2013) Acute mountain sickness: prediction and treatment during climbing expeditions (academic dissertation)

Khanna K, Mishra KP, Ganju L, Kumar B, Singh SB (2018) High-altitude-induced alterations in gut-immune axis: a review. Int Rev Immunol 37(2):119–126

Kramer AF, Coyne JT, Strayer DL (1993) Cognitive function at high altitude. Hum Factors 35(2): 329–344

Kumar R, Jain V, Kushwah N, Dheer A, Mishra KP, Prasad D, Singh SB (2021) HDAC inhibition prevents hypobaric hypoxia-induced spatial memory impairment through PI3K/GSK3β/CREB pathway. J Cell Physiol 236(9):6754–6771

Kushwah N, Jain V, Kadam M, Kumar R, Dheer A, Prasad D, Kumar B, Khan N (2021) *Ginkgo biloba* L. prevents hypobaric hypoxia–induced spatial memory deficit through small conductance calcium-activated potassium channel inhibition: the role of ERK/CaMKII/CREB signaling. Front Pharmacol 12:669701

Lieberman P, Morey A, Hochstadt J, Larson M, Mather S (2005) Mount Everest: a space analogue for speech monitoring of cognitive deficits and stress. Aviat Space Environ Med 76(6):198–207

Luks AM, Swenson ER, Bärtsch P (2017) Acute high-altitude sickness. Eur Respir Rev 26(143): 160096

Maiti P, Singh SB, Sharma AK, Muthuraju S, Banerjee PK, Ilavazhagan G (2006) Hypobaric hypoxia induces oxidative stress in rat brain. Neurochem Int 49(8):709–716

Paralikar SJ (2012) High altitude pulmonary edema-clinical features, pathophysiology, prevention and treatment. Indian J Occup Environ Med 16(2):59

Regard M, Oelz O, Brugger P, Landis T (1989) Persistent cognitive impairment in climbers after repeated exposure to extreme altitude. Neurology 39(2):210–210

Ruan YW, Ling GY, Zhang JL, Xu ZC (2003) Apoptosis in the adult striatum after transient forebrain ischemia and the effects of ischemic severity. Brain Res 982(2):228–240

Schild L, Reinheckel T, Reiser M, Horn TF, Wolf G, Augustin W (2003) Nitric oxide produced in rat liver mitochondria causes oxidative stress and impairment of respiration after transient hypoxia. FASEB J 17(15):2194–2201

Sharma NK, Sethy NK, Meena RM, Ilavazhagan G, Das M, Bhargava K (2011) Activity-dependent neuroprotective protein (ADNP)-derived peptide (NAP) ameliorates hypobaric hypoxia induced oxidative stress in rat brain. Peptides 32:1217–1224

Sharma NK, Sethy NK, Bhargava K (2013) Comparative proteome analysis reveals differential regulation of glycolytic and antioxidant enzymes in cortex and hippocampus exposed to short-term hypobaric hypoxia. J Proteomics 79C:277–298

Stieg PE, Sathi S, Warach S, Le DA, Lipton SA (1999) Neuroprotection by the NMDA receptor-associated open-channel blocker memantine in a photothrombotic model of cerebral focal ischemia in neonatal rat. Eur J Pharmacol 375(1–3):115–120

Stobdan T, Karar J, Pasha MQ (2008) High altitude adaptation: genetic perspectives. High Alt Med Biol 9(2):140–147

Storz JF, Cheviron ZA (2021) Physiological genomics of adaptation to high-altitude hypoxia. Annu Rev Anim Biosci 9:149–171

Sutton JR (1992) Mountain sickness. Neurol Clin 10(4):1015–1030

Turner RE, Gatterer H, Falla M, Lawley JS (2021) High altitude cerebral edema-its own entity or end-stage acute mountain sickness. J Appl Physiol 131:313–325

Virués-Ortega J, Buela-Casal G, Garrido E, Alcázar B (2004) Neuropsychological functioning associated with high-altitude exposure. Neuropsychol Rev 14(4):197–224

West RL (1996) An application of prefrontal cortex function theory to cognitive ageing. Psychol Bull 120(2):272

West T, Atzeva M, Holtzman DM (2007) Pomegranate polyphenols and resveratrol protect the neonatal brain against hypoxic-ischemic injury. Dev Neurosci 29(4–5):363–372

Wilson MH, Newman S, Imray CH (2009) The cerebral effects of ascent to high altitudes. Lancet Neurol 8(2):175–191

Woods P, Alcock J (2021) High-altitude pulmonary edema. Evol Med Public Health 9(1):118–119

Wright AD, Brearey SP, Imray CHE (2008) High hopes at high altitudes: pharmacotherapy for acute mountain sickness and high-altitude cerebral and pulmonary oedema. Expert Opin Pharmacother 9(1):119–127

Yarnell PR, Heit J, Hackett PH (2000) High-altitude cerebral edema (HACE): the Denver/front range experience. Semin Neurol 20(2):209–218

Zhang J, Yan X, Shi J, Gong Q, Weng X, Liu Y (2010) Structural modifications of the brain in acclimatization to high-altitude. PLoS One 5(7):e11449

Zubieta-Calleja G, Zubieta-DeUrioste N (2021) Acute mountain sickness, high altitude pulmonary edema, and high altitude cerebral edema: a view from the high Andes. Respir Physiol Neurobiol 287:103628

High Altitude Sickness: Environmental Stressor and Altered Physiological Response

2

Vartika, Sunanda Joshi, Monika Choudhary, Sameer Suresh Bhagyawant, and Nidhi Srivastava

Abstract

By the end of the nineteenth century, humans have been fascinated by and drawn to mountains and mostly peaks had been climbed by numerous mountaineers. Some early mountaineers reported experiencing the symptoms now known as mountain sickness or "Hypobaric hypoxia". Even today, many questions about the precise mechanism of altitude remain unanswered. Pulmonary and cerebral syndromes are a common clinical condition in unacclimatized people after a physical transformation at higher altitudes with millions of people travelling to high altitude each year. Acute mountain sickness (AMS) is a common syndrome that typically manifests itself several hours after the rise. Loss of appetite, dizziness, retching, sleep disruption, exhaustion, and drowsiness are classic causes of headache. At higher altitudes, sleep disturbances may become more severe, mental performance may suffer, and weight loss may occur. Certain medication including Acetazolamide and Ibuprofen can reduce the risk of developing AMS and can quickly relieve symptoms for the management of the potentially fatal conditions. The goal of this chapter is to provide a perspective for managing high altitude sicknesses and recommending the unacclimatized high altitude traveller by integrating a dialogue of risk factors associated,

Vartika · S. Joshi · M. Choudhary
Department of Bioscience and Biotechnology, Banasthali Vidyapith, Banasthali, Rajasthan, India

S. S. Bhagyawant
School of Studies in Biotechnology, Jiwaji University, Gwalior, Madhya Pradesh, India

N. Srivastava (✉)
Department of Biotechnology, National Institute of Pharmaceutical Education and Research (NIPER), Raebareli, Uttar Pradesh, India
e-mail: nidhi1.srivastava@niperraebareli.edu.in

© The Author(s), under exclusive license to Springer Nature Singapore Pte Ltd. 2022
N. K. Sharma, A. Arya (eds.), *High Altitude Sickness – Solutions from Genomics, Proteomics and Antioxidant Interventions*,
https://doi.org/10.1007/978-981-19-1008-1_2

preventive measures, and potential treatments with a description of the rudimentary physiological functions to elevation hypoxia.

Keywords

High altitude · Hypobaric hypoxia · Acute mountain sickness · High altitude pulmonary oedema · Cerebral oedema · Sleep disruption

2.1   Introduction

The definition of altitude varies depending on the context and it refers to the location's elevation above ocean level or the perpendiculars between being an imaginary plane (reference datum) and a spot or entity. They are linked to atmospheric pressure as analysed by an altimeter and as height increases; atmospheric pressure declines due to Planet's force of gravity or just because the molecules in the air reduces as height increases, attempting to make the air less intense. Altitude could be further categorized depending on its length from sea level (Fig. 2.1).

High altitudes (HAs) are areas mostly on planet's surface (or in its atmosphere) that are considerably higher than sea level. High altitudes are also sometimes described as beginning at 8000 ft above sea level. Wanting to stay at high elevations

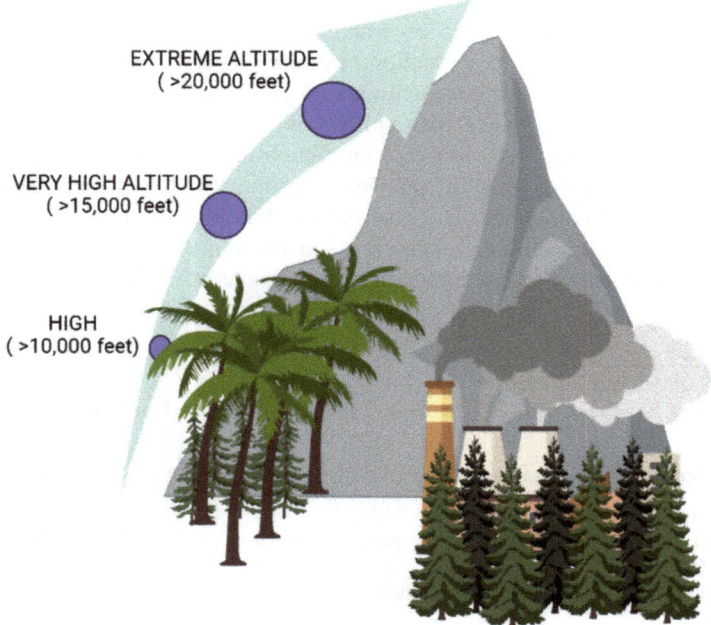

Fig. 2.1 Pictorial representation of different altitudes

for a longer length of time, either as part of an adventure or for the defence of one's country usually includes residing in a stressful situation (Chamberlin 2015).

2.1.1   High Altitudes

The high elevations of mountain ranges have always fascinated and inspired people as they are quiet and calm. The mountain ranges encouraged various psychics, intellectuals, and philosophers and provided the hostile climate; as well as untapped resources to the climbers that tempt hominids to sojourn in provinces even at higher reaches of mountain ranges. The framework of a substantial civilization in the Himalayan region (3500–4500 m) is an outstanding example of effective altitude alterations (Wu 2001; Aldenderfer 2003; Yuan et al. 2007; Shi et al. 2008; Zhao et al. 2009) (Fig. 2.2).

Hypoxia is by far the most substantial and strange aspect of higher latitudes. This is a condition have low level of oxygen accessibility that has a detrimental effect on both memory and cognitive function. The high altitudes of mountains have always fascinated and inspired people but cold, dehydration, increased exposure to solar radiation, malnourishment, and hypobaric hypoxia are certain obstacles faced on daily existence at high elevations (Bouverot 1985). This chapter introduces readers to the concepts of high altitudes, the global scenario, and hypoxia. One such foundational understanding is needed for a greater understanding of high altitude pathophysiology and strategic planning. Elevated stress factors usually involve enhanced ultraviolet radiation, hypobaric, hypoxemia, dangerous weather conditions, failure to maintain sufficient hygiene habits, cluttered living spaces, and alienation from adequate healthcare (Basnyat and Murdoch 2003) (Fig. 2.3).

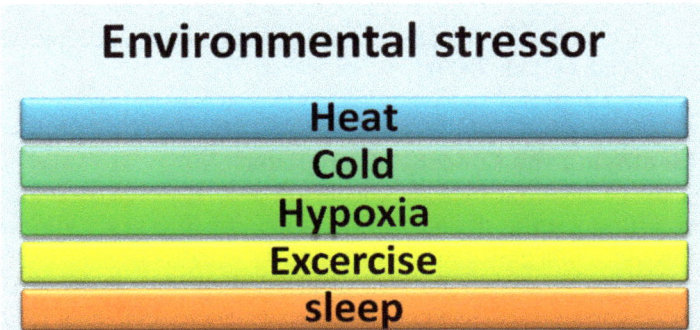

Fig. 2.2 Various factors related to high altitude

Fig. 2.3 Causes of Hypoxia

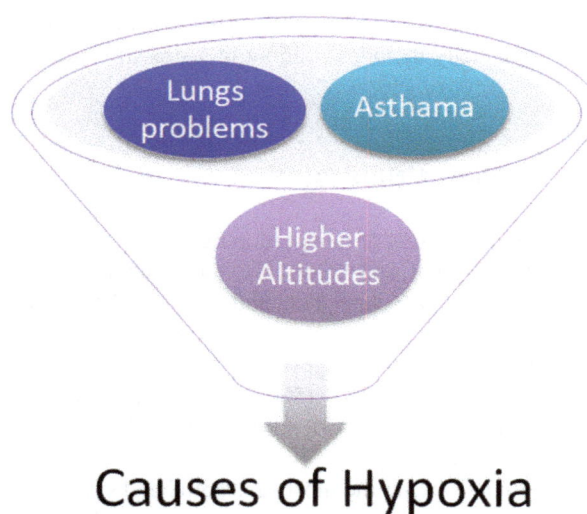

Causes of Hypoxia

2.1.2   Environment at High Altitudes and Defining "Hypoxia"

The environment surrounding mountain contains numerous factors that really are unfamiliar to straightforward denizens, causing a change in physiology and equilibrium. The most critical and intrinsic aspects about such an ecosystem are hypoxia (lack of oxygen), chilly (the temperature drops), airflow (cold, causing chill), enhanced radiation from the sun, ozone concentration, lack of plants and animals, and separation from human society. Independently, these components are potent psychophysical causes of stress. They pose a shocking barrier to human adaptability and survival capacity when they coexist in varying amounts in different geographical areas and weather conditions. However, when attempting to define all other stress factors, hypoxia remains the primary consideration. It occurs at high altitudes due to lower atmospheric pressure than at sea level. This issue arose as a result of two factors: gravity and the heat content, which causes compounds to tumble off each other (Hackett and Coach 2001). Because oxygen is more important to life than water and food, high altitude hypoxia has a long-term impact on people who live in this ecosystem (Julian et al. 2009).

2.1.2.1 Identifying the Concept of "Hypoxia"

According to the book on hypoxic conditions, they demarcated anoxemia as a nonexistence of oxygen in the blood (Van Liere and Stickney 1963). Numerous different researchers suggested that body continues to suffer from an oxygen deprivation even when there is no shortfall in blood. Literature revels that the term "anoxia" is described as a situation of oxygen starvation throughout the body and the research also uncovered three different types of anoxia: anoxic, anaemic, and stagnant (Barcroft 1920) (Fig. 2.4).

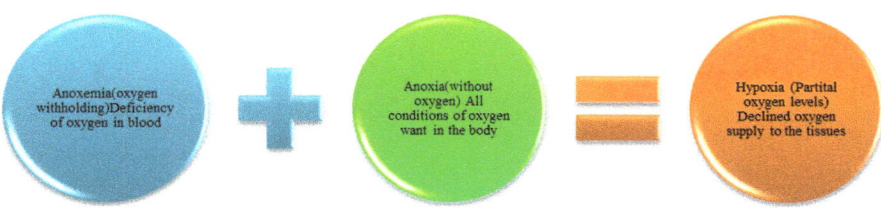

Fig. 2.4 Foundation of word "Hypoxia"

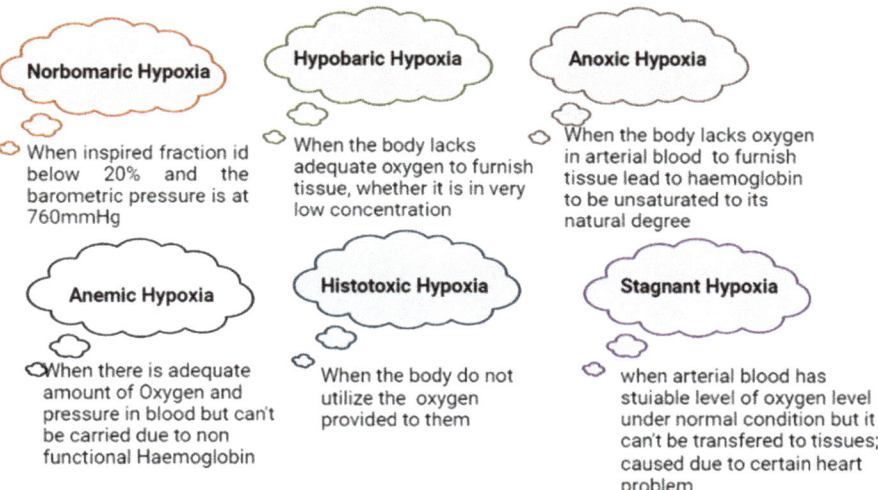

Fig. 2.5 Various types of hypoxia

2.1.2.2 Types of Hypoxia

Numerous researchers characterize various types of hypoxia based on the type of oxygen deprivation (Van Liere and Stickney 1963; Millet et al. 2012) (Fig. 2.5).

2.1.2.3 Symptoms of High Altitude Hypoxia

Whenever there is an inadequacy of biochemical adaptation to elevation hypoxic conditions, symptoms such as shortness of breath, physically and mentally tiredness, accelerated heart rate, interrupted sleep, as well as aches and pains that have been aggravated by physical exercise could perhaps arise. There could also be some gastrointestinal challenges and, in the some instances, a significant weight loss over a few days or even weeks. Sometimes in situations, people may also experience respiratory failure, stomach pain, as well as vomiting. Under very rare instances, at elevations (~ or <4500 m), there'll be a significant reduction in eyesight, abdominal discomfort, and gum bleeding. Though other people are supposed to be familiarized to elevated sickness, some may only encounter minimal side effects which can be overcome with altitude training. That many of those who appear to be used to

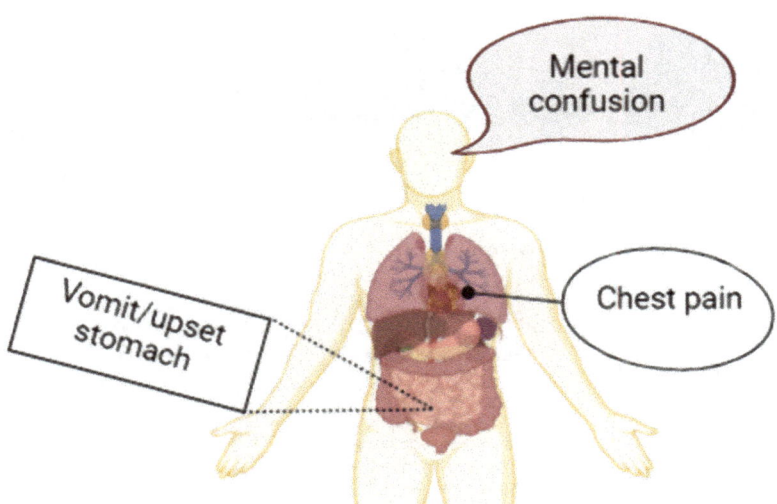

Fig. 2.6 Symptoms of Hypoxia

mountain illness would be unable to make adjustments and treat complications (Rahn and Otis 1949; Monge et al. 1992). Biological and physical factors influence the impacts of increased hypoxia; at around 1500 m, some clinical manifestations might well be observed. At that high altitude, there is almost no impacts at rest; however, hypoxia consequences could become apparent during physical exercise. The impacts of hypoxia are felt throughout both rest and physical exercise between 6000 and 10,000 ft. The physical responses become much more evident and inevitable above 2500–3000 m, and the physiological restrictions of human acceptance to high altitude hypoxia appear to be reached at 30,000–35,000 ft (Frisancho 2013) (Fig. 2.6).

2.1.3 Ultraviolet Radiation

Higher intensity of ultraviolet radiation (UVR) is because high altitude settings are at close proximity to the sun and thinner atmosphere. There is abundant proof of a causal relation between UVR exposure and the occurrence of melanoma and non-melanoma skin cancers, heightened incidence of eye problems, and skin ageing and diseases related to vitamin D (World Health Organization (WHO) 2006). Even so, the foreseeably negative effects of incredibly harsh biomedical research notifications underestimating the pernicious influence of UV rays stimulation had already recently been described (Lucas et al. 2006). Elevated contact with UV rays seems to defend alongside certain types of cancer and improved survival degrees in patients with melanoma (early-stage) appear to be related to continued sun exposure (Hughes et al. 2004; Berwick et al. 2005; Smedby et al. 2005). The development of breast, prostate, and colon cancers is a risk factor due to vitamin D deficiency,

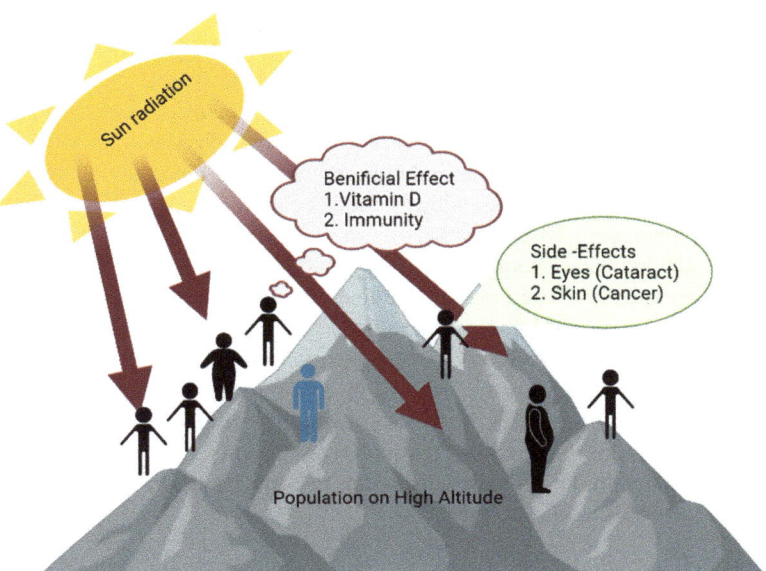

Fig. 2.7 Effect of UV rays on High altitude

though the evidence is not convincing (Lucas et al. 2006). The worker at high altitude with long winters can lead to vitamin D deficiency if they are not encouraged to get enough UVR exposure (Rivera-Ch et al. 2008) (Fig. 2.7).

2.1.4   Cold

A series of intriguing experiments on animals have revealed that simultaneous exposure to cold and hypoxia results in an up surged Pulmonary Artery Pressure (PAP) compared to the animals that are exposed to hypoxia or cold distinctly. Similarly individual present at high altitude settings have surprisingly more strength on the physiological and pathological responses mounted to overcome compared to ones at low lands (Chauca and Bligh 2003). In this regard, it is interesting to note that attenuated ventilation has been demonstrated in cold conditions rather than warm conditions, and the attenuation appears to be greater when hypoxia is added to cold (Mortola and Frappell 2000). It was proposed that the importance of keeping blood circulation in winter during hypoxia be balanced against reducing heat loss, which also limits the airflow spike. Living creatures can help combat the freezing soundscape through warmer wear and domestic heating. They could, however, ignore the consequences of temperature variations, which would've been especially important in high altitude climatic conditions where life quality appears to be low and under nutrition is widespread. The health consequences of chronic continuous or infrequent cold exposure and hypoxia are unknown. It is also uncertain whether enhanced erythropoietic activity observed in response to hypoxia is influenced by concurrent cold conditions. These and other physiological and biochemical responses in cold/hypoxic conditions are undoubtedly important, and they may pave the way for future human studies, particularly on the risk of developing clinical conditions associated with acute and chronic exposure. Physicians working in high altitude settings, as well as parents and children, who live there, are more concerned about whether cold in hypoxic conditions predisposes to acute respiratory infections, particularly pneumonia. Most hypoxic environments have low ambient temperatures and have made increased frequency of respiratory infections, in young children and the elderly particularly. The impact of ambient temperature on the development of respiratory infections and pneumonia has been hotly debated in the medical literature. Steadily increasing person-to-person interaction and indoor pollution, instead of lower temperature exposure, will be a more sensible explanation for such increased incidence of respiratory diseases during the cold season (Rivera-Ch et al. 2008). Acute respiratory infections, particularly in children, have also been shown to exacerbate high altitude hypoxia, increasing the risk of death (Duke et al. 2001; Niermeyer 2003; Huicho 2007).

2.2   Physiological Changes at High Altitude

Humans experience physiological stress as a result of the high altitude (2000–3000 m) environment. The body adapts and changes in order for the oxygen transport system to compensate for the hypoxia and maintain an adequate tissue oxygen level to support metabolism (Hurtado 1971). Hyperventilation, increases in oxidative enzymes, changes in oxygen affinity of haemoglobin, hypoxic pulmonary vasoconstriction, polycythaemia, and increased concentration of capillaries are all classic responses to the chances that occur at high alltitudes (Hurtado 1971; Frisancho 1993). An increase in altitude causes a drop in pressure, which reduces atmospheric oxygen pressure and causes hypobaric hypoxia, which affects the physiological functioning of the body (Frisancho 1993; West et al. 2019) (Fig. 2.8).

The most significant physiologic effects of high altitude on the human body are on mental performance and sleep.

- Mental Performance (MP)
  Non-acclimatized individuals experience physiological disturbances, as well as adverse effects in neuropsychological efficiency at cruising altitude (<3000 m) (West 2004). It is well documented that HA exposure could have a detrimental effect psychological efficiency, and emotional states, which include anxiousness; that obviously depends on the flight level touched, the rate of incline, and the hours invested at HA (Lowe et al. 2007; Dykiert et al. 2010). Although many people working at an elevation of 4000 m have an overwhelming amount of arithmetic inaccuracies, a shorter attention span, and more emotional exhaustion. Exposure to nearly 2500 m altitude has an adverse influence on auditory and visual responsiveness, as well as short attention span (San et al. 2013).

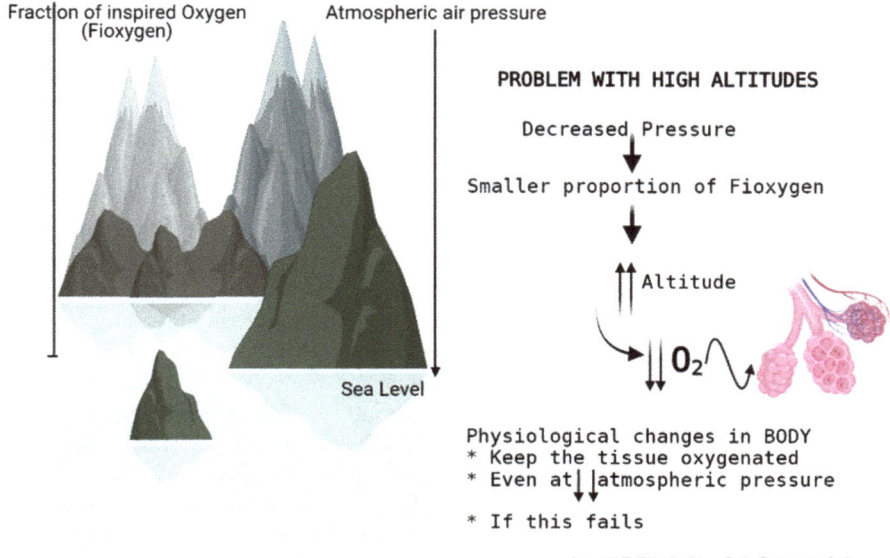

Fig. 2.8 Physiological changes at high altitude and altitude sickness

The cellular and molecular underlying mechanisms impeded MP throughout hypoxia remain a mystery. The central nervous system customarily uses 20% of the body's total O_2; that has been completely used only for oxidative metabolism. Adjustments in redox homeostasis, modifications in calcium absorption, transformations in neurotransmission metabolic rate, and disability of synaptic feature have all been proposed as pathways for nerve cell function dysfunction throughout hypoxic conditions. (West 2004; Lowe et al. 2007; Dykiert et al. 2010). Cardiorespiratory mechanisms have an influence on MP and may result in a condition called as an organic brain disorder while attempting to climb to HA (Sevre et al. 2001). Environmental influences, pollution levels, workout, and physical characteristics throughout elevation ascent could indeed have a detrimental effect on MP (Sevre et al. 2001; West 2004).

• Sleep
The decreased oxygen level of the body tends to cause respirations uncertainty at elevation, with intervals of deep and rapid breathing referred to as high altitude periodic breathing (PB). It can happen to healthy individuals at altitudes above 6000 ft. It may cause sleep problems, frequent awakenings, and a perception being out of breath (Lombardi et al. 2013). Hypoxia lessened sleep quality, and effectiveness, along with rapid eye movement that likely to result in depressed episode, rage, and exhaustion, which exacerbated focus, visual, working memory, ability to concentrate, cognitive processes, inhibitory control, and psychological response time. After 24 h, altered sleep rhythms could indeed influence mental state and cognitive function (de Aquino Lemos et al. 2012). People at HA quite

often wake up, have outbursts, and do not feel calm in the early hours during the day and they suffer from drowsiness (Weil 2004). The known cause is probably periodic breathing (PB), which happens in the majority of the population at elevations above 4000 m (Cingi et al. 2010; Johnson et al. 2010). Males are much more affected by nocturnal periodic breathing at high elevation than women. Only at the highest reached altitude did females begin to exhibit a significant number of central sleep apneas in this survey. Gender differences in the apnea-hypopnea index were still observed after 10 days at 5400 m, as were discrepancies in respiratory cycle length after chronic exposures (Latshang et al. 2012). PB entails alternating periods of shallow and deep breathing. Four to five deep inhalations are typically accompanied by a bunch of really breathing or even a full pause in respiration that is defined as apnea (Johnson et al. 2010). During nap at HA, carbon dioxide concentration in the body could indeed decline to critical levels that can turn off the desire to inhale. Ventilation is resumed only when the body senses another such drop in oxygen levels. PB is believed to be caused by control system disruption caused by the hypoxic drive or the response to CO_2 (Cingi et al. 2010; Johnson et al. 2010). The elevation sleep disorder was caused primarily by pulmonary disruptions caused by the physiologic ventilatory challenge of acute ascent, in which activation by hypoxia alternated with suppression by hypocapnic alkalosis (Groves et al. 1987; Anholm et al. 1992; Weil 2004). Hiking tall is most often suggested by knowledgeable hikers and mountaineers, but sleeping alleviates such problems. The chilly breezes, loud or stinky tent mates, and long distance travel all can interfere with sleep. According to the findings, quality of sleep is at first; that is dysfunctional in healthful climbers trying to climb quickly to high elevation, but enhances with acclimatization in affiliation with enhanced oxygenation, while regular intervals respiration continues to exist. As a result, elevated troubled sleep show up to be triggered predominantly by hypoxia instead of by periodic breathing (Reite et al. 1975; Nussbaumer-Ochsner et al. 2012).

2.2.1   Acclimatization to High Altitude

Acclimatization refers to a series of beneficial changes that occur in the body as a result of a response or adaptive change, and it significantly improves human tolerance to high altitude. Hyperventilation and polycythemia are the most important physiological responses during acclimatization (Fig. 2.9).

2.2.1.1 Pulmonary Ventilation or Hyperventilation

The most unique feature of acclimatisation is increased respiratory duration and level, which leads to an increase alveolar ventilation (West 2004). In response to low O_2 concentrations in arterial blood, the hypoxic ventilatory response (HVR) of peripheral chemoreceptors, primarily the carotid bodies located just above the bifurcation of the common carotid artery, is activated (Howald and Hoppeler 2003; Calbet and Lundby 2009). The HVR is the reflex reaction to hypoxic

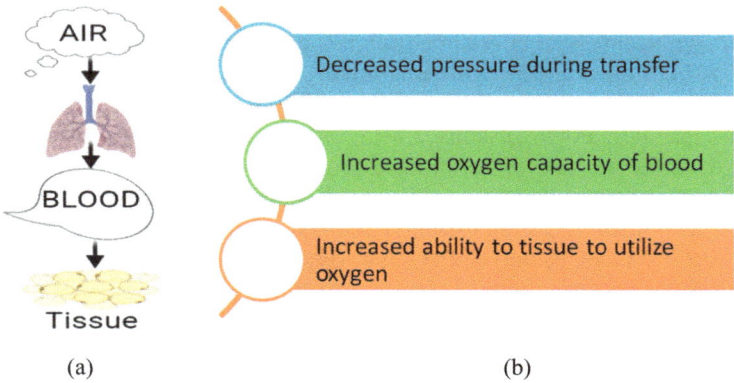

(a)   (b)

Fig. 2.9 Pictorial representation of (**a**) Acclimatization to high altitude (**b**) Pathways involved in delivery of oxygen to tissue

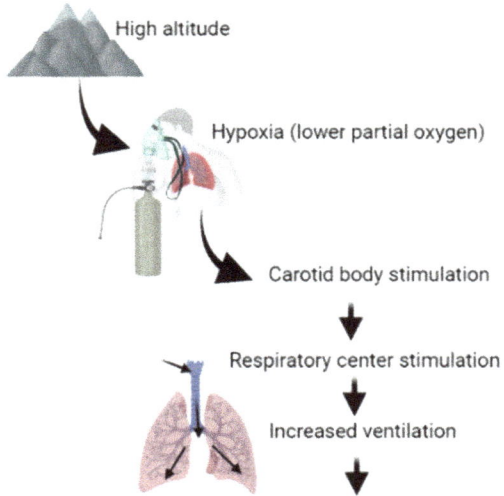

Fig. 2.10 Schematic presentation of Hyperventilation

stimulation from carotid-body chemo receptors. The time-dependent increase in HVR that occurs after hours to weeks of hypoxic exposure is referred to as hypoxic ventilation acclimatization. Hyperventilation increases the partial pressure of alveolar (PPA) and PaO_2 while decreasing PPA and arterial CO_2 (West 2004; Calbet and Lundby 2009) (Fig. 2.10).

2.2.1.2 Polycythaemia

Another significant effect of acclimatization is polycythaemia, a blood disorder also known as high altitude polycythaemia (HAPC), which is a chronic high altitude disease caused by an abnormal increase in red blood cell mass and thus high blood oxygen capacities. Polycythaemia, on the other hand, develops slowly. It takes several days for the rate of erythrocyte production to increase. HAPC primarily causes an increase in blood viscosity, resulting in microcirculatory and immune response disruptions such as vascular thrombosis, extensive organ damage, and sleep disorders (Hocking and Golde 1989) (Fig. 2.11).

2.2.2 Affinity of Haemoglobin for Oxygen at Sea Level and At High Altitudes

In comparison to their lowland relatives, vertebrate taxa native to high altitude environments have higher haemoglobin (Hb)–O_2 affinities. To cope with the low ambient PO_2 at high altitude, air-breathing vertebrates' blood O_2 transport capacity must be increased in order to sustain O_2 flux to tissue mitochondria in support of aerobic ATP synthesis. These modifications supplement physiological changes in other convective and diffusive steps of the O_2 transport pathway. Although increased Hb–O_2 affinity helps to protect arterial O_2 saturation in the face of environmental hypoxia, it can also impede O_2 unloading in the systemic circulation. As a result, vertebrates at high altitude face the physiological challenge of balancing O_2 loading in the pulmonary capillaries and O_2 unloading in the tissue capillaries (Storz 2016).

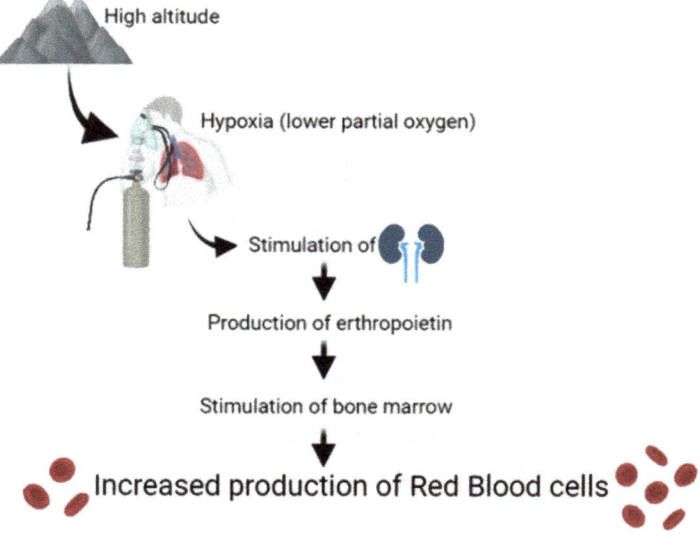

Fig. 2.11 Schematic presentation of Polycythaemia

2.2.3 High Altitude and Oxidative Stress

As several RONS systems are generated in the body, aerobic metabolism produces the activation of reactive oxygen and nitrogen (RONS). Indeed, in a number of known and unknown physiological and pathophysiological processes RONS are natural and physiologic modulators of a cellular redox environment that signal control factors. An increase in the supply of oxygen leads to an increase in the production of mitochondrial ROS. Furthermore, it has been suggested that 1–2% of the oxygen attempting to enter the mitochondrion is released as ROS. Hypoxia, on the other hand, appears to result in reductive stress, which leads to increased ROS production by the mitochondrial electron transport system (Mohanraj et al. 1998). During hypoxia, less O_2 is available for cytochrome oxidase to reduce to H_2O, resulting in an accumulation of reducing equivalents within the mitochondrial respiratory sequence. This build-up is known as reductive stress, and it results in the formation of ROS via the auto-oxidation of one or more mitochondrial complexes, such as the ubiquinone–ubiquinol redox couple. High altitude reduces the efficiency of the antioxidant system and, as a result of increased RONS production, can cause oxidative damage to macromolecules. Physical activity can exacerbate the effects of high altitude and increase the oxidative stress associated with it (Dosek et al. 2007). Exposure to high altitude reduces oxygen pressure and increases the formation of reactive oxygen and nitrogen species (RONS), which is frequently associated with increased oxidative damage to lipids, proteins, and DNA. The antioxidant enzyme system appears to be less active and effective at high altitude. Furthermore, during high altitude exposure, several RONS generating sources are activated, including the mitochondrial electron transport chain, xanthine oxidase, and nitric oxide synthase (NO). High altitude physical activity can exacerbate oxidative stress (Bakonyi and Radak 2004) (Fig. 2.12).

2.3 Conclusion

Today, high elevation territories around are constantly being investigated and became permanent housing for around 2% of the global population. There is an inhospitable place of excessive hypoxia and hypothermia, low barometric pressure, and intense ultraviolet radiation exposure. The important role in adaptation to high altitude is the respiratory system, which is essential for tissue oxygenation. The basic physiologic mechanism of high altitude diseases is low PO_2 gas is inspired as a result of decreased barometric pressure. In healthy people, the most significant consequences of ascending to high altitude are divided into three categories: reduced maximal oxygen consumption, impaired mental performance, and sleep disturbances. The effects from hyperventilation caused by hypoxic stimulation of peripheral chemoreceptors are greatly reduced by acclimation, high altitude, the main feature. But a common misunderstanding about acclimation is that the body is restored to near-normal standards, a serious mistake. People living close to sea level have increasing demands for working at high altitudes and recent significant

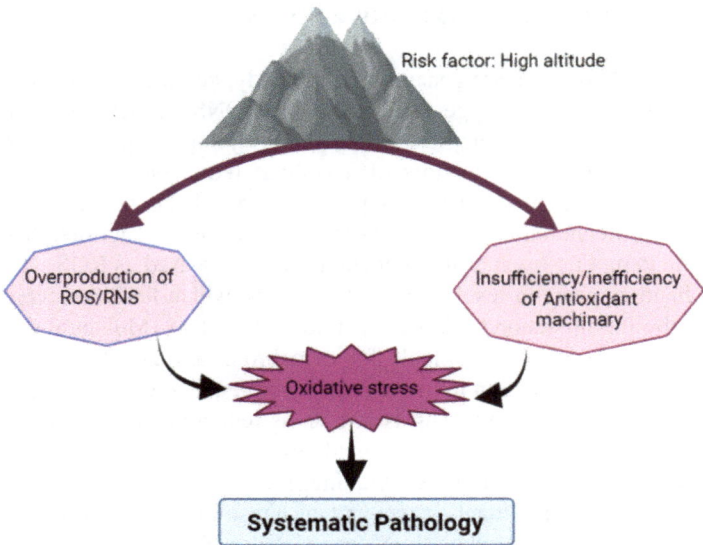

Fig. 2.12 Risk factors related with high altitude and oxidative stress

improvements, enrichment of room air oxygen, productivity increase, fatigue reduction, and sleep improvement. The misadjustment to high altitude conditions causes various acute, subacute and chronic diseases, such as acute mountain sickness, emoemas of the high altitude of the lungs, and emoemas of the cerebrasis of high altitude, which are far more serious and have an influence on the respiratory system. High altitude cerebral oedema can also be fatal, but the mechanism is unknown. All three conditions respond favourably to immediate descent.

Prophylactic medication and strict adherence to acclimate schedules significantly reduce the risk that lowlanders will experience high altitude-related acute disease. Many physiological benefits will be gained for long-term residents who successfully acclimatize and adapt. Higher regions of the world in all history inspired many people to conquer nature by reaching higher elevations. The high altitude regions of the world not only pose physiological challenges, but also provide an environment to enhance humanity.

References

Aldenderfer MS (2003) Moving up in the world: archaeologists seek to understand how and when people came to occupy the Andean and Tibetan plateaus. Am Sci 91(6):542–549

Anholm JD, Powles AC, Downey R, Houston CS, Sutton JR, Bonnet MH, Cymerman A (1992) Operation Everest II: arterial oxygen saturation and sleep at extreme simulated altitude. Am Rev Respir Dis 145(4 Pt 1):817–826

Bakonyi T, Radak Z (2004) High altitude and free radicals. J Sports Sci Med 3(2):64

Barcroft J (1920) On anoxaemia. Lancet 2(4):485

Basnyat B, Murdoch DR (2003) High-altitude illness. Lancet 361(9373):1967–1974
Berwick M, Armstrong BK, Ben-Porat L, Fine J, Kricker A, Eberle C, Barnhill R (2005) Sun exposure and mortality from melanoma. J Natl Cancer Inst 97(3):195–199
Bouverot P (1985) The respiratory gas exchange system and energy metabolism under altitude hypoxia. In: Adaptation to altitude-hypoxia in vertebrates. Springer, Berlin, Heidelberg, pp 19–34
Calbet JA, Lundby C (2009) Air to muscle O2 delivery during exercise at altitude. High Alt Med Biol 10(2):123–134
Chamberlin R (2015) Survival medicine tips techniques & secrets. https://www.prepperwebsite. com/. A survival medicine & medical preparedness blog sharing tips, techniques and secrets for building the perfect first aid kit and using it to treat injuries and illnesses preppers encounter during disasters. Accessed 23 Jul 2021
Chauca D, Bligh J (2003) The interactive stresses of hypoxia and cold at high altitude. In: Hipoxia: Investigaciones Básicas y Clínicas. Homenaje a Carlos Monge Cassinelli. Travaux de l'Institut Français d'Études Andines, Tome, vol 76, pp 103–110
Cingi C, Erkan AN, Rettinger G (2010) Ear, nose, and throat effects of high altitude. Eur Arch Otorhinolaryngol 267(3):467–471
de Aquino Lemos V, Antunes HKM, dos Santos RVT, Lira FS, Tufik S, de Mello MT (2012) High altitude exposure impairs sleep patterns, mood, and cognitive functions. Psychophysiology 49(9):1298–1306
Dosek A, Ohno H, Acs Z, Taylor AW, Radak Z (2007) High altitude and oxidative stress. Respir Physiol Neurobiol 158(2–3):128–131
Duke T, Mgone J, Frank D (2001) Hypoxaemia in children with severe pneumonia in Papua New Guinea [oxygen therapy in children]. Int J Tuberc Lung Dis 5(6):511–519
Dykiert D, Hall D, van Gemeren N, Benson R, Der G, Starr JM, Deary IJ (2010) The effects of high altitude on choice reaction time mean and intra-individual variability: results of the Edinburgh altitude research expedition of 2008. Neuropsychology 24(3):391
Frisancho AR (1993) Human adaptation and accommodation. University of Michigan Press, Ann Arbor, MI
Frisancho AR (2013) Developmental functional adaptation to high altitude. Am J Hum Biol 25(2): 151–168
Groves BM, Reeves JT, Sutton JR, Wagner PD, Cymerman ALLEN, Malconian MK, Rock PB, Young PM, Houston CS (1987) Operation Everest II: elevated high-altitude pulmonary resistance unresponsive to oxygen. J Appl Physiol 63(2):521–530
Hackett PH, Coach RC (2001) High-altitude illness. N Engl J Med 345:107–114
Hocking WG, Golde DW (1989) Polycythemia: evaluation and management. Blood Rev 3(1): 59–65
Howald H, Hoppeler H (2003) Performing at extreme altitude: muscle cellular and subcellular adaptations. Eur J Appl Physiol 90(3):360–364
Hughes AM, Armstrong BK, Vajdic CM, Turner J, Grulich AE, Fritschi L, Milliken S, Kaldor J, Benke G, Kricker A (2004) Sun exposure may protect against non-Hodgkin lymphoma: a case-control study. Int J Cancer 112(5):865–871
Huicho L (2007) Postnatal cardiopulmonary adaptations to high altitude. *Respir Physiol Neurobiol 158*(2–3):190–203
Hurtado A (1971) The influence of high altitude on physiology. In: High altitude physiology: cardiac and respiratory aspects. Churchill Livingstone, London, pp 3–13
Johnson PL, Edwards N, Burgess KR, Sullivan CE (2010) Sleep architecture changes during a trek from 1400 to 5000 m in the Nepal Himalaya. J Sleep Res 19(1p2):148–156
Julian CG, Wilson MJ, Moore LG (2009) Evolutionary adaptation to high altitude: a view from in utero. Am J Hum Biol 21(5):614–622
Latshang TD, Bloch KE, Lynm C, Livingston EH (2012) Traveling to high altitude when you have sleep apnea. JAMA 308(22):2418–2418

Lombardi C, Meriggi P, Agostoni P, Faini A, Bilo G, Revera M, Caldara G, Di Rienzo M, Castiglioni P, Maurizio B, Gregorini F (2013) High-altitude hypoxia and periodic breathing during sleep: gender-related differences. J Sleep Res 22(3):322–330

Lowe M, Harris W, Kane RL, Banderet L, Levinson D, Reeves D (2007) Neuropsychological assessment in extreme environments. Arch Clin Neuropsychol 22(Suppl_1):S89–S99

Lucas RM, Repacholi MH, McMichael AJ (2006) Is the current public health message on UV exposure correct? Bull World Health Organ 84:485–491

Millet GP, Faiss R, Pialoux V (2012) Point: counterpoint: hypobaric hypoxia induces/does not induce different responses from normobaric hypoxia. J Appl Physiol 112(10):1783–1784

Mohanraj P, Merola AJ, Wright VP, Clanton TL (1998) Antioxidants protect rat diaphragmatic muscle function under hypoxic conditions. J Appl Physiol 84(6):1960–1966

Monge C, Arregui A, Leon-Velarde F (1992) Pathophysiology and epidemiology of chronic mountain sickness. Int J Sports Med 13(S 1):S79–S81

Mortola JP, Frappell PB (2000) Ventilatory responses to changes in temperature in mammals and other vertebrates. Annu Rev Physiol 62(1):847–874

Niermeyer S (2003) Cardiopulmonary transition in the high altitude infant. High Alt Med Biol 4(2): 225–239

Nussbaumer-Ochsner Y, Ursprung J, Siebenmann C, Maggiorini M, Bloch KE (2012) Effect of short-term acclimatization to high altitude on sleep and nocturnal breathing. Sleep 35(3): 419–423

Rahn H, Otis AB (1949) Man's respiratory response during and after acclimatization to high altitude. Am J Physiol Legacy Content 157(3):445–462

Reite M, Jackson D, Cahoon RL, Weil JV (1975) Sleep physiology at high altitude. Electroencephalogr Clin Neurophysiol 38(5):463–471

Rivera-Ch M, Castillo A, Huicho L (2008) Hypoxia and other environmental factors at high altitude. Int J Environ Health 2(1):92–106

San T, Polat S, Cingi C, Eskiizmir G, Oghan F, Cakir B (2013) Effects of high altitude on sleep and respiratory system and their adaptations. Sci World J 2013:241569

Sevre K, Bendz B, Hankø E, Nakstad AR, Hauge A, Kåsin JI, Lefrandt JD, Smit AJ, Eide I, Rostrup M (2001) Reduced autonomic activity during stepwise exposure to high altitude. Acta Physiol Scand 173(4):409–417

Shi H, Zhong H, Peng Y, Dong YL, Qi XB, Zhang F, Liu LF, Tan SJ, Ma RZ, Xiao CJ, Wells RS (2008) Y chromosome evidence of earliest modern human settlement in East Asia and multiple origins of Tibetan and Japanese populations. BMC Biol 6(1):1–10

Smedby KE, Hjalgrim H, Melbye M, Torrång A, Rostgaard K, Munksgaard L, Adami J, Hansen M, Porwit-MacDonald A, Jensen BA, Roos G (2005) Ultraviolet radiation exposure and risk of malignant lymphomas. J Natl Cancer Inst 97(3):199–209

Storz JF (2016) Hemoglobin–oxygen affinity in high-altitude vertebrates: is there evidence for an adaptive trend? J Exp Biol 219(20):3190–3203

Van Liere EJ, Stickney JC (1963) Hypoxia. In: Library of Congress catalog card number: 63–16722. The University of Chicago Press, Chicago and London

Weil JV (2004) Sleep at high altitude. High Alt Med Biol 5(2):180–189

West JB (2004) The physiologic basis of high-altitude diseases. Ann Intern Med 141(10):789–800

West J, Schoene R, Luks A, Milledge J (2019) High altitude medicine and physiology 5E. CRC, Boca Raton, FL

World Health Organization (WHO) (2006) Global burden of disease from solar ultraviolet radiation. WHO, Geneva. http://www.who.int/uv/health/solaruvradfull_180706.pdf. Accessed 21 Jul 2021

Wu T (2001) The Qinghai–Tibetan plateau: how high do Tibetans live? High Alt Med Biol 2(4): 489–499

Yuan B, Huang W, Zhang D (2007) New evidence for human occupation of the northern Tibetan plateau, China during the late Pleistocene. Chin Sci Bull 52(19):2675–2679

Zhao M, Kong QP, Wang HW, Peng MS, Xie XD, Wang WZ, Jiayang, Duan JG, Cai MC, Zhao
    SN, Cidanpingcuo, Tu YQ, Wu SF, Yao YG, Bandelt HJ, Zhang YP (2009) Mitochondrial
    genome evidence reveals successful late Paleolithic settlement on the Tibetan plateau. Proc Natl
    Acad Sci U S A 106:21230–21235

High Altitude Related Diseases: Milder Effects, HACE, HAPE, and Effect on Various Organ Systems

3

Sneha Singh and Mairaj Ahmed Ansari

Abstract

Proteomics has emerged as an excellent biomarker discovery tool in recent years owing to the rapid emergence and growth of high throughput proteomics approaches and advancements in mass spectrometry methods especially in terms of mass analysers and resolution. Moreover, plasma and serum have always been a choice of diagnostic fluid since historical times and have always been a preferred diagnostic sample by clinicians. Although plasma is disproportionate in terms of composition as some proteins such as albumin and globulins are highly abundant, while others are found in traces. The emergence of methods to deplete the high abundance of proteins has led to a remarkable refinement in plasma or serum composition for its diagnostic abilities. Plasma proteins are always considered the gold standard for several metabolic changes induced by diseases or altered physiology. High altitude related physiological changes have also been evaluated recently over the proteomic scale and seemingly represent a potential source for exploring biomarkers for such physiological changes. This chapter describes the hypobaric hypoxia-induced changes in the plasma and serum proteome and potential biomarkers which have enormous commercial and clinical potential.

Keywords

High altitude · Diseases · AMS · HAPE · HACE

S. Singh · M. A. Ansari (✉)
Department of Biotechnology, School of Chemical and Life Sciences, Jamia Hamdard, New Delhi, India

37

N. K. Sharma, A. Arya (eds.), *High Altitude Sickness – Solutions from Genomics, Proteomics and Antioxidant Interventions*,
https://doi.org/10.1007/978-981-19-1008-1_3

3.1   Introduction

Mountains occupy nearly a quarter of the Earth's land area (Kapos et al. 2000) and remain a dwelling place to 140 million people across the globe. These statistics not only include remote, economically disadvantaged groups of communities such as Northern India, Tibet, and Nepal, but also several urban centres and adjacent areas to hills such as valleys, including some megacities such as Santigo, Chile, and Mexico (Meybeck et al. 2001). Apart from permanent residence, high altitudes are also visited by millions of individuals for recreational activities such as skiing, hiking, mountaineering and even peak climbing, performance budding training of athletes; religious pilgrimages and in some cases as military support to safeguard the national territories on the military front. Altitudes have always been foreseen as one of the most beautiful creations of nature, and always admired. But the darker side also exists, particularly, for the people ascending to an altitude beyond 2500 m at a steep rate from sea level.

In such conditions high altitude poses its own set of risks and challenges to the human physique and psyche, resulting in several sickness symptoms and worst-case scenarios, death. The major cause for these ill effects that have been revealed in many scientific studies is a reduction in the partial pressure of air (hypobaric hypoxia) causing altered cardiopulmonary responses associated with pathophysiology thereof. Although some of the high altitude regions are not accessible to common people yet, others are accessible and frequently visited by travel enthusiasts and inhabitants. High altitudes of the greater Himalayas in the north of India, especially in the Leh district are some of the frequently visited and well-inhabited high altitudes places. Not only due to local inhabitants but also due to strategic situations, these areas have some of the highest often visited places. Moreover, the Leh-Nubra highway is also connected by the highest motorable road with the highest mountain top existing at a staggering altitude of 17,982 feet (5481 m), commonly known as Khardung La Pass (Fig. 3.1). The Himalayas, particularly due to strategic reasons, have many such inhabited high altitude stations and roads in India, China, Nepal, Pakistan and Afghanistan.

Although researchers have not developed a consensus that a distinct altitude boundary persists between a high altitude and low altitude, yet, from the scientific evidence on clinical symptoms, generally, high altitude is considered above 1500 m above mean sea level (MSL). Several high altitude experts have stratified high altitude into three levels, first, high altitude at an altitude between 1500 m and 3500 m; second, very high altitude from 3500 to 5500 m; and third, extreme altitude beyond 5500 m of altitude (Bakonyi and Radak 2004). The highest altitude noted on the lithosphere is 8848 m (Mount Everest) and has been an attraction for mountaineers. Several other mountains in the Himalayas, Swiss Alps, and Andes have altitudes reaching 6096 m (20,000 ft), and numerous tourist attractions in the altitude range of 3000–5000 m. As one moves to a higher altitude the rate of air becomes sparse due to lower atmospheric pressure and thereby causing decreased partial pressure of oxygen. The rate of decrease in atmospheric pressure is exponential (Fig. 3.2) (Peacock 1998). Although the most impacting environmental effect of

Fig. 3.1 A view of the highest motorable road in the world at Khardung La Pass, (**a**) Motorable road across the valley. (**b** and **c**) A view of hill-top at Khardung La pass, (**d**) Signboard for tourists indicating the hill-top altitude of 17,982 feet or 5481 m. (Photo—Author)

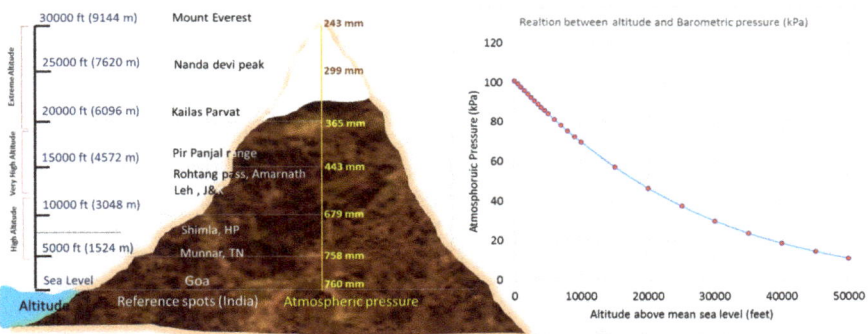

Fig. 3.2 High altitude and changes in relative atmospheric pressure, Altitude above 5000 ft is considered high altitude, 11,500 ft–18,000 ft. as very high altitude and 18,000- beyond is considered as extreme altitude (some reference tourist locations are mentioned from Indian terrain). The barometric pressure drops exponentially with increase in altitude (right panel)

high altitudes is hypoxia, yet the low temperatures and dry environment in some arid areas cannot be ignored while considering the overall impact on human physiology.

At low atmospheric pressure, partial pressure (pO_2) of oxygen also decreases, which adversely affects human physiology (Rick 1995) leading to high altitude illness. The rate of ascent determines the severity of illness. The first clinical symptoms usually appear within 2 days of reaching the high altitude. Among the most commonly reported symptoms are fatigue, nausea and vomiting, headache, breathlessness, inability to sleep, and occasional swelling of the face and limbs (Hultgren 1996). Rapid ascent to high altitude is often known to be associated with some additional physiological changes such as breathlessness, drowsiness, loss of appetite, loss of sleep, lassitude, and irritability to further ascend to higher altitude. These aforementioned symptoms are commonly referred to as acute mountain sickness (AMS in short) or high altitude sickness. To develop AMS at an altitude above 5000 m, a period of a few hours is required. Nevertheless, exposure at altitudes as low as 2500 m (8000 ft) have also been sensitive enough for few people and led to the development of AMS. The development of AMS at these altitudes may be significantly increased after a long-haul flight. It is generally suggested that a person on arrival at a high altitude should take proper rest for at least 24 h and should avoid any physical activity and may extend the period of rest if mild symptoms appear (Wilson et al. 2009). While the milder effects remain limited to the afore-mentioned acute symptoms which are relieved with rest and prolonged stay, the chronic and more severe forms of altitude sickness might impact vital organs such as the lungs and brain to a great extent while other vital organs such as the heart, liver, and kidney as a secondary manifestation of the effect on brain and heart. The most severe form of mountain sickness that affects the lung is known as high altitude pulmonary oedema or HAPE in short, while the most severe and occasionally lethal effect of altitude sickness on the brain is known to be high altitude cerebral oedema or HACE in short. During the International Hypoxia Symposium in 1991, a consensus was developed over the proposal of Peter Hackett and Oswald Oelz (Hartig and Hackett 1992) to define and quantify the various altitude illnesses. This later turned into a draft in 1992 and evaluation criteria for high altitude sickness commonly known as Lake Louise criteria. The scoring system includes severity on a scale of zero to three for the five common symptoms related to high altitude, i.e. headache, gastrointestinal upset, and fatigue/weakness, dizziness/light T headedness, and sleep disturbance. A total score ≥ 3, in the presence of a headache, was decided to be considered as a diagnostic feature for AMS. An updated version of the consensus redrafted in the year 2018 is provided in Table 3.1.

3.2 Epidemiology of High Altitude Sickness

If one considers the epidemiology of high altitude sickness, scanty and sparse evidence and data is available. There is a lack of suitable global data accessible to clinicians despite enormous reporting and efforts made by local authorities and clinical teams. The incidences of high altitude sickness are known to increase with ascending altitude. While high altitude sickness is reportedly seldom at an altitude below 2500 m, the percentage of travellers facing difficulties in acclimatization above 3000 m is approximately 75%. Moreover, travellers with a clinical history of high altitude sickness are more susceptible than those who have surpassed similar trips in the past (Prince et al. 2021) The susceptibility of high altitude sickness is exaggerated by pre-existing ailments such as asthma, anaemia, hepatic or renal dysfunctions. Although it is difficult to keep the global record of high altitude sickness, yet the emerging networking and informatics tools have allowed the collection to be integrated. One such integration of data on high altitude illness

Table 3.1 2018 Lake Louise Acute Mountain Sickness Score (After Roach et al. 2018)

Headache	Dizziness/light-headedness
[a]None at all	[a]No dizziness/light-headedness
[b]A mild headache	[b]Mild dizziness/light-headedness
[c]Moderate headache	[c]Moderate dizziness/light-headedness
[d]Severe headache, incapacitating	[d]Severe dizziness/light-headedness, incapacitating
Gastrointestinal symptoms	AMS clinical functional score
[a]Good appetite	[a]Not at all
[b]Poor appetite or nausea	[b]Symptoms present, but no change in activity
[c]Moderate nausea or vomiting	[c]Symptoms forced me to stop the ascent or to go down
[d]Severe nausea and vomiting, incapacitating	[d]Had to be evacuated to a lower altitude
Fatigue and/or weakness	
[a]Not tired or weak	
[b]Mild fatigue/weakness	
[c]Moderate fatigue/weakness	
[d]Severe fatigue/weakness, incapacitating	

[a]Score of zero or no effect
[b]Score 1
[c]Score 2
[d]Score 3. (Based on original Lake Louis Criteria)

was performed by Hackett and coworkers in 2004 that reported global sampling of high altitude sickness. It is known that a correlation exists between altitude and the number of AMS cases.

For example, in one study by Maggiorini et al., reports of AMS incidences have been described, where they have indicated about 9% at an altitude of 2850 m in the swiss alps, while 13% at 3050 m, and above 3650 m it was nearly 36% (Maggiorini et al. 1990). Besides AMS, HAPE, a much severe version of high altitude sickness, typically occurs at altitudes over 3000 m, but it has been recorded at a lower altitude threshold of 1400 m.

In a study, containing a brief epidemiological profile of HAPE, ND Menon, has been described the prevalence of HAPE with clinical evidence, estimating it in a range from 0.2% at an altitude of 4550 m to approximately 15% in sojourns ascending to an altitude of 3500 m or above especially via flight (Menon 1965). Compared to HAPE, HACE is less life-threatening in terms of epidemiology as evidently, it is much less common than HAPE and ACE. The incidences of HACE are known to be less than 1–2% and observed only above the altitude of 4000 m. HACE is often, but not always, preceded by the development of HAPE. Some efforts towards epidemiological data collections have been made by organizations such as altitude.org as a part of the apex project led by Dr. J. Kenneth Baillie (https://www.altitude.org/), where they have started collecting the global HAPE data and also host valuable information related to altitude sickness. The statistics of deaths occurring at high altitude is also sparsely available and no integrated global data is yet available.

The primary reason for this non-follow-up of the patients and also, secondary reasons of deaths such as accidents or avalanche. Most of the travellers which travel to high altitude for personal tourism are unregistered to a database except those who are on special mountain expeditions, such as Mount Everest. Death statistics of the Everest summit from 1921 to 2006 provided in a study by Firth et al., showed that the mortality rate among mountaineers above base camp was 1.3%, which included both traumatic and non-traumatic deaths and even some deaths were reported as missing individuals. A number of altitude sickness symptoms such as profound fatigue, ataxia, and cognitive changes were also reported in non-traumatic deaths (Firth et al. 2008).

3.3   Mechanisms of High Altitude Sickness

Onset and progression of high altitude sickness may be categorized into three steps, (a) hypoxia stimulus, (b) sensing, and (c) molecular changes. Hypoxia is the primary stimulus since the initiation of the pathogenesis of acute mountain sickness, although symptoms become visible 6–7 h after the exposure, but worsen with increasing altitude (Savonitto et al. 1992) and relieved by normalizing the inspired PO_2 (Wu et al. 2007). These stimuli are sensed by cellular oxygen sensing systems and thereby evoking a compensatory response. Two major compensatory events are evoked, increased pressure (mechanical) and increased permeability (molecular). Increased permeability is regulated by potent vasodilators such as nitric oxide (NO) while pressure is increased due to hyperventilation response. These two changes adversely affect the cardiopulmonary and cerebral systems by increased capillary pressure resulting in accumulation of fluids in the brain and lung causing high altitude cerebral oedema (HACE) and high altitude pulmonary oedema (HAPE),respectively.

One of the acute changes during the high altitude related hypoxia at the physiological level is hypoxic pulmonary vasoconstriction or HPV, which can be considered as an adaptive physiological response. A few individuals with strong hypoxic pulmonary vasoconstriction (HPV) capacity rarely develop high altitude pulmonary oedema. Hypoxia is known to directly influence the vascular tone of the pulmonary vessels, as well as elevates the systemic resistance across vessels. This change is reportedly attributed to the peripheral chemoreceptors. Furthermore, the cross-talk between the ventilatory response of hypoxia chemoreceptor-mediated reflex on the cardiopulmonary physiology results in exacerbating the condition. Peripheral chemoreceptors activation leads to increased ventilation and hence alkalosis as well as sympathetic activation of the autonomic nervous systems thereby increasing heart rate, cardiac output, myocardial contraction velocity, and blood pressure. Alkalosis is not compensated by usual renal compensation mechanisms above the altitude of 3000–4000 m and thus leads to a rapid fall in serum potassium which can be more problematic for diuretic patients. Cold and exercise and further add the sympathetic activation and aggravate the symptoms. In fact, cold changes in renal function at high altitudes are also observed as a direct effect of hypoxia on the kidney

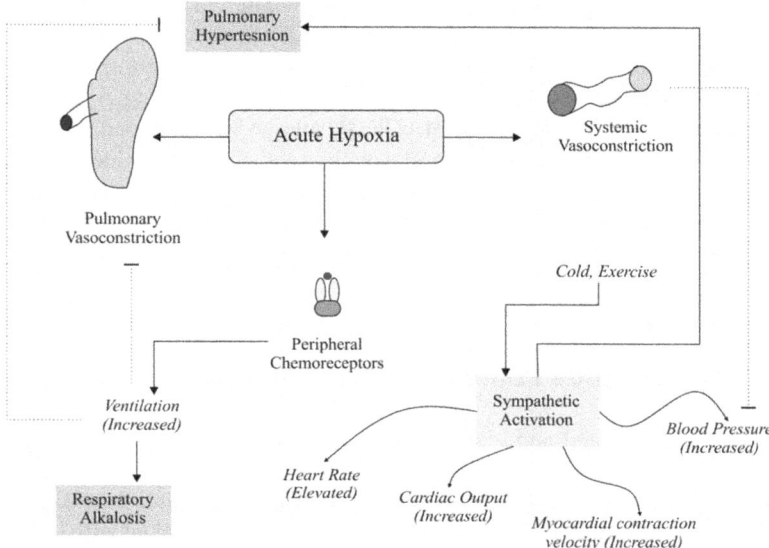

Fig. 3.3 Outline Mechanism of origin of high altitude sickness. Acute hypoxia at altitudes results in either direct impact on the tone of blood vessels by causing either pulmonary vasoconstriction or systemic vasodilation and secondly, sensory activity of peripheral chemoreceptors leads to increased ventilation or sympathetic activation of various cardiac functions. Renal functions are affected due to increased alkalosis and inability to perform compensation for respiratory alkalosis, while exaggerating the sympathetic activation and cardiac functions

especially as a renal compensation of hypoxia-induced alkalosis. As a part of this compensatory activity, urine output, sodium excretion, diuresis and natriuresis, elevated potassium and bicarbonate excretion. These changes are also activated by chemoreceptor based signalling as discussed above. These responses vary almost by about tenfold in the first 24 to 48 at high altitudes. Besides this, changes in renin, angiotensin, aldosterone, atrial natriuretic peptide, vasopressin, also impart their signalling effect, especially via nitric oxide messenger and further to hypoxia-inducible factor (Fig. 3.3).

3.4   Milder Effects of High Altitude (AMS)

Acute mountain sickness is initial milder symptoms initiated at the rapid ascent to a high altitude, which include headache, dizziness, tiredness, loss of appetite, and shortness of breath. AMS is not the same as high altitude cerebral oedema (HACE) and therefore must not be clinically confused as it is not known to be associated with prominent neurological perturbation. AMS is associated with much milder symptoms in contrast to much worse conditions as HAPE or HACE, which usually comes on between 24 and 72 h after a gain in altitudes. As the worsening of symptoms can happen quickly after AMS, provoking factors such as running, exertion, high-physical activity, and smoking should be strictly avoided at extreme altitudes (Hackett et al. 1988; Willmann et al. 2014). Also, alcohol use and sleeping pills are additional factors that increase the chances of developing Acute mountain sickness and worsen it to HAPE or HACE. Besides AMS, a related issue is called sub-acute mountain sickness (or SMAS), which is a distinct syndrome symptomized by congestive cardiac failure and observed in lowlanders while they have prolonged stay at extremely high altitudes. We have discussed this in more detail in later sections on cardiopulmonary aspects of high altitude sickness.

3.5   High Altitude Pulmonary Oedema (HAPE)

HAPE was first reported and elaborated by Ravenhill in 1913 (West 1996). High Altitude Pulmonary Oedema or HAPE develops usually within 2 to 4 days after reaching high altitude (>2500 m). It is characterized by dyspnea at rest, decreased tolerance to exercise and tightness of the chest. HAPE is also linked with tachycardia (slow heart rate), tachypnea, and dry cough often with lung crackles. Some other signs associated with HAPE are wheezing possibly due to accumulation of lung fluid, and cyanosis (Hartig and Hackett 1992). Fever and haemoptysis have occasional occurrences. The clinical correlation does not need to be homogenous in chest radiography as it can vary across individuals, most commonly appearing as patchy alveolar infiltrates, the spread of which increases with severity (Marticorena and Hultgren 1979; Vock et al. 1991). Although the exact mechanism of HAPE is not well understood, some suggested mechanisms include an exaggerated hypoxic pulmonary vasoconstriction with an abnormal increase in the pulmonary artery (Hultgren 1996; Penaloza and Sime 1969; Allemann et al. 2000). This further leads to the irregular distribution of vasoconstriction with regional over perfusion (Zhao et al. 2001) and increased capillary pressure (Maggiorini et al. 2001) finally causing trans-microvascular fluid leakage (Schoene et al. 1988; Swenson and Maggiorini 2002). It is not completely understood that genetic predisposition is linked to the increased endothelin production in HAPE susceptible patients with lower nitric oxide production, yet several other studies have reported nitric oxide modulations and their correlation with HAPE. A study at Author's lab demonstrated that dietary nitrite attenuates oxidative stress and activates antioxidant genes in rat heart during hypobaric hypoxia (Singh et al. 2012). HAPE is differentiated from AMS on the grounds of exhaustion, hypothermia, hyponatraemia, migraine, dehydration, infection, carbon monoxide poisoning, drug and alcohol intoxication, hypoglycaemia or severe hyperglycaemia, transient ischaemic attack or stroke and acute psychosis, possibly related to intake of corticosteroids.

3.6   High Altitude Cerebral Oedema (HACE)

HACE is a potentially fatal neurological condition that is known to develop after hours to days in high altitude travellers post AMS or HAPE. HACE is considered as a terminal clinical manifestation of AMS, it is not well understood, however, that HACE is always preceded by AMS, or HAPE and can occur independently without any clinical evidence of HAPE. Clinically it is correlated, by ataxia and hallucinations, and accumulation of fluid in the brain. Coma and death are eventual results if HACE is left untreated. The clinical diagnosis of HACE is based on its cardinal features such as a change in consciousness and ataxia (Sutton 1992). Mental status changes may include irrational behaviour that rapidly progresses to lethargy, obtundation, and coma. Other physical indications of HACE useful in clinical diagnosis are bleeding retinal veins (retinal haemorrhage), palsies of cranial nerve palsies, and cognitive deficits. The exaggerated form of these clinical conditions leads to brain herniation which causes death in HACE patients (Hultgren 1996; Yarnell et al. 2000; Hackett 1999). The increased intracranial pressure is one of the signs that can also be imaged using MRI. The underlying cause behind the oedema of the brain is the elevation of intracranial pressure, which is the overperfusion of blood vessels in response to global vasodilation mediated by nitric oxide and other related molecular events. However, there is also a possible role of other factors such as radicals and radical-induced adaptive signalling causing the pathology of HACE.

3.7   High Altitude Related Cardiac Perturbations

Although it is well-known that pulmonary oedema that appears during high altitude sickness is primarily of non-cardiac origin. However, cardiac effects are mostly retrospective in the occurrence of high altitude sickness. It is reported that cardiac or pulmonary diseases are at a much higher risk of high altitude sickness. Nevertheless, compared to patients with coronary heart disease, hypertension or bronchial asthma have lesser ill effects, while patients with chronic obstructive pulmonary disease (COPD), pulmonary hypertension or interstitial pulmonary disease might be impacted to a greater extent (Fischer 2004). As we have discussed above in the mechanism of high altitude sickness, one of the key signals that impact the heart is the activation of sympathetic systems due to chemoreception of acute hypoxia. A special condition of acute mountain sickness that is known to impact the heart to a great extent is known as adult sub-acute mountain sickness (SAMS). Poduval in his clinical assessment of SAMS has reported that individuals stationed above 5000 m when evaluated for the signs of congestive cardiac failure using simple chest X-ray and electroencephalography. Exertional dyspnea and bilateral pedal oedema were observed as the most common sign in experimental patients. Moreover, some patients also showed signs of deep venous thrombosis (Poduval 2000).

The risk of myocardial ischemia is enhanced by an elevation in cardiac work during the first few days at altitude but reduces significantly as the cone relieves the cardiac work possibly by rest and lower physical activity.. Exercise-mediated coronary flow reserves are known to be significantly reduced at 2500 m in patients with coronary disease leading to coronary spasm mediated by sympathetic activation and alkalosis. The synergistic effect of hypoxia also with exercise is a tremendous stressor for the heart and pulmonary system. However, the cases of deaths occurring due to exercise at altitude are limited. Alkalosis of blood mediated by hyperventilation poses an indirect threat to the cardiac system, especially for diuretics. Patients with uncontrolled hypertension which are administered diuretics may need a rescheduling and revised dosing at high altitude. Also, they should avoid visiting altitude until their blood pressure is controlled (Bärtsch and Gibbs 2007).

3.8 High Altitude Related Perturbations in Muscles

Not much has been explored about the changes in muscles by acute high altitude exposure. However, it is likely to be highly diverse based on the duration of stay, altitude, degree of physical activity, and prior histology of muscles. Nevertheless, on the biochemical grounds, hypoxia at the tissue level (still needs to be completely validated for muscles) is a possible cause of switching of muscle to glycolytic mode. Some studies confirmed the upregulation of glycolytic enzymes in muscle post-hypoxic exposure. It is also evident that the muscles of high altitude dwellers have elevated expression of glycolytic enzymes in comparison to lowlanders (Rosser and Hochachka 1993, Green et al. 1998). Reynafarje B in their study in the early 1960s has also reported a significantly higher level of myoglobin in high altitude dwellers compared to lowlanders (Reynafarje 1962). Yet, these differences between high altitude natives and lowlanders are not a correlative measure of the effects of high altitude and acute exposure of hypobaric hypoxia on human muscles. Moreover, hypoxic cachexia, which is a well-described anatomic and physiological perturbation in hypoxia, remains poorly understood at the mechanistic level. Additionally, chronic hypoxia is also known to affect the fibre composition and enzyme activities in muscles primarily by shifting them from slow oxidative to fast glycolytic metabolism (Rosser and Hochachka 1993).

3.9 High Altitude Related Perturbations in Kidney

Physiologically, kidneys are voracious consumers of oxygen accounting for about 10% of the entire body's oxygen consumption despite only 0.5% of the mass. The presence of intra-renal hypoxia in several diseases such as diabetes is an indicator of the relation between hypoxia and kidney functions at the pathological level. There is evidence of the increased prevalence of nephropathy in diabetic patients staying at high altitudes. Laustsen et al. have recently elaborated this discussion and provided pieces of evidence on alteration of renal oxygen availability and explored the metabolism of pyruvate which was shifted to a higher side towards lactate and alanine rather than oxidative fate (Laustsen et al. 2014). More recently, it has been studies that humans living at high altitudes are at greater risk for developing high-altitude renal syndrome (HARS), which is symptomised by polycythemia, hyperuricemia, systemic hypertension and microalbuminuria (Hurtado et al. 2012). The primary reason for such renal perturbations at high altitudes is mainly attributed to acute systemic hypoxia as well as prolonged renal hypoperfusion.

In an attempt to study the physiology of the renal system under hypoxia, a study by Handitsch et al. leaves some impressive footprints by providing data on changes in important renal indicators after hypoxic exposure. They observed an initial decrease in glomerular filtration rate that became significant after 48 h which returned to normal after approximately a week Moreover, a sustained erythropoietin production was observed above an altitude of >2100 m which is suggestive of renal remodelling. As kidneys are also important organs performing the renal compensation for the hyperventilation and hypoxia-induced respiratory alkalosis, the sodium excretion rate and fractional excretion of sodium is recorded inconsistent (Haditsch et al. 2015).

3.10 High Altitude Related Perturbations in Liver

The liver is one of the busiest organs of the body in terms of metabolic pathways and also a site of control of many intermediate metabolic processes. After the brain and kidney, it is the largest consumer of oxygen, therefore hypoxia of various origins may affect its anatomy as well as physiology. Very limited research has been conducted and therefore scanty data is available on evidence of liver injury or dysfunction in individuals travelling to extreme altitudes of 5000 m or above. It is evident that, while the patients with liver transplants may not be affected by high altitude exposure, the travellers with cirrhosis require careful pre-travel evaluation. It is important to identify the predisposition to high altitude particularly in cirrhotic liver cases especially for mitigating the illness at those altitudes (Luks et al. 2008). While the book was being written, no evidence was available about the long-term risk of chronic liver disease, while staying at high altitudes or travelling to altitudes for a much longer period in months or years.

3.11 Conclusion

Hypoxia as it appears is a major physiological threat to a significant portion of the human population either travelling or dwelling at extreme altitudes. Despite decades of efforts the exact nature of physiological perturbations and pathological manifestations are not completely resolved at the mechanistic level. Albeit, research has provided suitable and safe regimens for altitude ascent in terms of adaptive measures and acclimatization methodologies. Rapid descent and oxygen supplementation have emerged as immediate remedies of high altitude sickness. Moreover, some prophylactic pharmaceuticals such as acetazolamide and dexamethasone have been proposed. Also, some therapeutic recommendations such as nifedipine, acetazolamide, sildenafil, and tadalafil have been established. A lot more evidence is warranted at the system level and organ level changes in physiology to prevent any unlikely mortality and relieve the discomfort on the rapid ascent to high altitude.

References

Allemann Y, Sartori C, Lepori M, Pierre S, Melot C, Naeije R, Scherrer U, Maggiorini M (2000) Echocardiographic and invasive measurements of pulmonary artery pressure correlate closely at high altitude. Am J Physiol Heart Circ Physiol 279(4):H2013–H2016

Bakonyi T, Radak Z (2004) High altitude and free radicals. J Sports Sci Med 3(2):64–69

Bärtsch P, Gibbs JS (2007) Effect of altitude on the heart and the lungs. Circulation 116(19): 2191–2202

Firth PG, Zheng H, Windsor JS, Sutherland AI, Imray CH, Moore GW, Semple JL, Roach RC, Salisbury RA (2008) Mortality on Mount Everest, 1921-2006: descriptive study. BMJ 337: a2654

Fischer R (2004) Reiseziel Hochgebirge. Höhenmedizinische Tipps für Gesunde und Risikopatienten [Visiting high altitudes—healthy persons and patients with risk diseases]. MMW Fortschr Med 146(8):33–34. 36–7. German

Green PS, Gridley KE, Simpkins JW (1998) Nuclear estrogen receptor-independent neuroprotection by estratrienes: a novel interaction with glutathione. Neuroscience 84(1):7–10

Hackett PH (1999) The cerebral etiology of high-altitude cerebral edema and acute mountain sickness. Wilderness Environ Med 10(2):97–109

Hackett PH, Roach RC, Schoene RB, Harrison GL, Mills WJ Jr (1988) Abnormal control of ventilation in high-altitude pulmonary edema. J Appl Physiol 64(3):1268–1272

Haditsch B, Roessler A, Krisper P, Frisch H, Hinghofer-Szalkay HG, Goswami N (2015) Volume regulation and renal function at high altitude across gender. PLoS One 10(3):e0118730

Hartig GS, Hackett PH (1992) Cerebral spinal fluid pressure and cerebral blood velocity in acute mountain sickness. In: Sutton JR, Coates G, Houston CS (eds) Hypoxia and mountain medicine. Queen City Press, Burlington, VT, pp 260–265

Hultgren HN (1996) High-altitude pulmonary edema: current concepts. Annu Rev Med 47:267–284. https://doi.org/10.1146/annurev.med.47.1.267

Hurtado A, Escudero E, Pando J, Sharma S, Johnson RJ (2012) Cardiovascular and renal effects of chronic exposure to high altitude. Nephrol Dial Transplant 27(Suppl 4):iv11–iv16. https://doi.org/10.1093/ndt/gfs427

Kapos V, Rhind J, Edwards M, Price MF, Ravilious C (2000) Developing a map of the world's mountain forests. In: Price MF, Butt N (eds) Forests in sustainable mountain development: a state-of-knowledge report for 2000. CABI, Wallingford, pp 4–9

Laustsen C, Lycke S, Palm F, Østergaard JA, Bibby BM, Nørregaard R, Flyvbjerg A, Pedersen M, Ardenkjaer-Larsen JH (2014) High altitude may alter oxygen availability and renal metabolism in diabetics as measured by hyperpolarized [1-(13)C]pyruvate magnetic resonance imaging. Kidney Int 86(1):67–74

Luks AM, Johnson RJ, Swenson ER (2008) Chronic kidney disease at high altitude. J Am Soc Nephrol 19(12):2262–2271

Maggiorini M, Bühler B, Walter M, Oelz O (1990) Prevalence of acute mountain sickness in the Swiss Alps. BMJ 301(6756):853–855. PMID: 2282425; PMCID: PMC1663993. https://doi.org/10.1136/bmj.301.6756.853

Maggiorini M, Melot C, Pierre S, Pfeiffer F, Greve I, Sartori C, Lepori M, Hauser M, Scherrer U, Naeije R (2001) High-altitude pulmonary edema is initially caused by an increase in capillary pressure. Circulation 103(16):2078–2083

Marticorena E, Hultgren HN (1979) Evaluation of therapeutic methods in high altitude pulmonary edema. Am J Cardiol 43(2):307–312

Menon ND (1965) High altitude pulmonary edema: a clinical study. N Engl J Med 8(273):66–73. https://www.altitude.org/. Accessed 20 Jul 2021

Meybeck M, Green P, Vörösmarty C (2001) A new typology for mountains and other relief classes: an application to global continental water resources and population distribution. Mount Res Dev 21(1):11

Peacock AJ (1998) ABC of oxygen: oxygen at high altitude. BMJ 317(7165):1063–1066

Penaloza D, Sime F (1969) Circulatory dynamics during high altitude pulmonary edema. Am J Cardiol 23(3):369–378

Poduval RG (2000) Adult subacute mountain sickness—a syndrome at extremes of high altitude. J Assoc Physicians India 48(5):511–513

Prince TS, Thurman J, Huebner K (2021) Acute Mountain sickness. In: StatPearls. StatPearls, Treasure Island, FL. PMID: 28613467

Reynafarje B (1962) Myoglobin content and enzymatic activity of muscle and altitude adaptation. J Appl Physiol 17:301–305

Rick C (1995) Outdoor action guide to high altitude: acclimatization and illnesses. Princeton University, Princeton, NJ

Roach RC et al (2018) The 2018 lake louise acute mountain sickness score. High Alt Med Biol 19 (1):4–6. https://doi.org/10.1089/ham.2017.0164

Rosser BW, Hochachka PW (1993) Metabolic capacity of muscle fibers from high-altitude natives. Eur J Appl Physiol Occup Physiol 67(6):513–517

Savonitto S, Cardellino GG, Giulio D, Pernpruner S, Roberta B, Milloz N, Colombo MDHP, Sardina M, Nassi G, Marraccini P (1992) Effects of acute exposure to altitude (3,460 m) on blood pressure response to dynamic and isometric exercise in men with systemic hypertension. Am J Cardiol 70(18):1493–1497

Schoene RB, Swenson ER, Pizzo CJ, Hackett PH, Roach RC, Mills WJ Jr, Henderson WR Jr, Martin TR (1988) The lung at high altitude: bronchoalveolar lavage in acute mountain sickness and pulmonary edema. J Appl Physiol 64(6):2605–2613

Singh M, Arya A, Kumar R, Bhargava K, Sethy NK (2012) Dietary nitrite attenuates oxidative stress and activates antioxidant genes in rat heart during hypobaric hypoxia. Nitric Oxide 26(1): 61–73. https://doi.org/10.1016/j.niox.2011.12.002

Sutton JR (1992) Mountain sickness. Neurol Clin 10(4):1015–1030

Swenson ER, Maggiorini M (2002) Salmeterol for the prevention of high-altitude pulmonary edema. N Engl J Med 347(16):1282–1285. author reply 1282-5

Vock P, Brutsche MH, Nanzer A, Bartsch P (1991) Variable radiomorphologic data of high altitude pulmonary edema. Features from 60 patients. Chest 100(5):1306–1311

West JB (1996) T.H. Ravenhill and his contributions to mountain sickness. J Appl Physiol (1985) 80(3):715–724

Willmann G, Gekeler F, Schommer K, Bärtsch P (2014 Jun) Update on high altitude cerebral edema including recent work on the eye. High Alt Med Biol 15(2):112–122

Wilson MH, Newman S, Imray CH (2009) The cerebral effects of ascent to high altitudes. Lancet Neurol 8(2):175–191. https://doi.org/10.1016/S1474-4422(09)70014-6

Wu TY, Ding SQ, Liu JL, Jia JH, Dai RC, Zhu DC, Liang BZ, Qi DT, Sun YF (2007) High-altitude gastrointestinal bleeding: an observation in Qinghai-Tibetan railroad construction workers on mountain Tanggula. World J Gastroenterol 13(5):774–780

Yarnell PR, Heit J, Hackett PH (2000) High-altitude cerebral edema (HACE): the Denver/front range experience. Semin Neurol 20(2):209–217

Zhao L, Mason NA, Morrell NW, Kojonazarov B, Sadykov A, Maripov A, Mirrakhimov MM, Aldashev A, Wilkins MR (2001) Sildenafil inhibits hypoxia-induced pulmonary hypertension. Circulation 104(4):424–428

High Altitude-Induced Oxidative Stress, Rheumatoid Arthritis, and Proteomic Alteration

4

Vikram Dalal, Vishakha Singh, and Sagarika Biswas

Abstract

Oxidative stress is the disruption in the equilibrium between the production of pro-oxidants such as peroxynitrite ($ONOO^-$), reactive oxygen species (ROS), reactive nitrogen species (RNS), and superoxide anion ($.O_2^-$), etc. and antioxidants such as catalase, dismutase, etc. Two major sources of oxidative stress are endogenous and exogenous. Enhanced hyperoxia or aerobic metabolism is assumed to have high levels of reactive oxygen and nitrogen species (RONS) that have a high ability to oxidative damage to the lipids, DNA, and protein. High altitude increased the generation of ROS or reduced antioxidants that are the major causes of oxidative damage to macromolecules. Excess supply of oxygen can increase mitochondrial ROS production. In hypoxia, the mitochondrial electron transport system causes the generation of ROS. Short- and long-term exposure to hypoxia can enhance the level of oxidative stress. Rheumatoid arthritis (RA) is a chronic autoimmune condition that can cause joint damage and deterioration of the bone. Oxidative stress in RA includes various causes such as the irregular distribution of adhesive molecules, autophagy induction, and synoviocyte resistance for apoptosis. Several hours of exposure to higher humidity and reduced pressure have a major worse effect on RA. In vivo, ex vivo, and in-cell oxidative stress can be calculated using various instruments such as flow cytometry, fluorescence microplate reader, and confocal

V. Dalal · V. Singh
Department of Biotechnology, Indian Institute of Technology Roorkee, Roorkee, Uttarakhand, India

S. Biswas (✉)
Department of Genomics and Molecular Medicine, CSIR-Institute of Genomics and Integrative Biology, New Delhi, Delhi, India
e-mail: sagarika.biswas@igib.res.in

microscopy, etc. Increased altitude is related to physiological responses to hypobaric hypoxia stress by an increment in oxygen supply and usage of oxygen for tissue via metabolic modulation.

Keywords

Oxidative stress · Rheumatoid arthritis · Hypoxia · High altitude · Reactive species · Proteome

Abbreviations

5-LO	5-Lipoxygenase
AGE	Advanced glycation end product
AOPP	Advanced oxidation of protein products
CL-HPLC	Chemiluminescence-high performance liquid chromatography
CPT1B	Carnitine palmitoyltransferase 1
CT	3-Chlorotyrosine
CYP2E1	Cytochrome P450 2E1
DAF-2DA	Diaminofluorescein diacetate
DCF	Dichlorofluorescein
DCFDA	Dichlorofluorescein diacetate
DHR	Dihydrorhodamine 123
DPPP	Diphenyl-1-pyrenylphosphine
ESR	Electron spin resonance
FAO	Fatty acid oxidation
GSH	Glutathione
HIF	Hypoxia Induce factor
IgG	Immunoglobulin
IR	Ionization Radiation
LDH	Lactate Dehydrogenase
MBL	Mannose-Binding Lectin
MDA	Malondialdehyde
NO	Nitric oxide
NQO	Quinine oxidoreductase
OA	Osteoarthritis
PC	Protein carbonyls
PCOOH	Phosphatidylcholine
PEOOH	Phosphatidylethanolamine
PPARα	Peroxisome proliferator-activated receptor α
RA	Rheumatoid Arthritis
ROS	Reactive Oxygen Species
SOD	Superoxide dismutase
TBARS	Thiobarbituric acid reactive substances
TLRs	Toll-like receptors
TRX	Thioredoxin

4.1   Oxidative Stress

The disturbance in the formation of free radicals and its detoxification by the biological system leads to oxidative stress. Disturbance of the production of antioxidants such as catalase, scavengers, and dismutase, etc. and oxidants such as hydroxyl radicals (.OH), reactive nitrogen species (RNS), superoxide anion (.O_2^-), reactive oxygen species (ROS), and peroxynitrite ($ONOO^-$), etc. are referred as oxidative stress. Reactive oxygen species referred as intermediate products of biochemical reactions such as neutrophil-mediated phagocytosis, mitochondrial respiration, and cytochrome P450, etc.

4.1.1   Oxidants

Oxidative stress is a leading cause of the oxidative process and apoptosis, which further leads to cell death. Details of different oxidants are mentioned in Table 4.1. In oxidative phosphorylation, mitochondrial active oxygen leakage is the main source of the production of reactive oxygen radicals. Several redox-active flavoproteins may act as an important factor in oxidant development. Superoxide is produced by various enzymes such as xanthine oxidase, Nicotinamide adenine dinucleotide oxidase (NADPH), and cytochrome P450, etc. Four major endogenous sources of oxidants are: aerobic respiration reduced the O_2 and generate .O_2, .OH, and H_2O_2; phagocytosis of bacteria or virus-infected cells generate the O_2^-, nitric oxide (NO), OCl, and H_2O_2; H_2O_2 produced by peroxisome and animal cytochrome P450 generate intermediate products which can damage DNA.

ROS and RNS are known to be involved in the pathogenesis of schizophrenia and Alzheimer's diseases. Oxidative stress can cause hyperoxia, tissue injury, diabetes, and age-related development of cancer. Oxidants are mutagenic and cause DNA

Table 4.1 Different oxidants and their properties

Superoxide anion (.O_2^-)	One electron reduction of O_2 forms the .O_2^-. It is produced as an intermediate in various auto-oxidative reactions and electron transport chain
Hydrogen peroxide (H_2O_2)	Dismutation of O_2^- forms a two-electron reduction state ROS named as H_2O_2. It can easily cross the plasma membrane due to its lipid solubility
Hydroxyl radical (.OH)	It is produced by the decomposition of peroxynitrite and Fenton reaction. It is three electron reduction state radical and extremely reactive
Organic hydroperoxide (ROOH)	Cellular components like nucleobases and lipids form the ROOH
Peroxynitrite ($ONOO^-$)	Reaction between O^{-2} and NO˙ produce the peroxynitrite

damage directly or indirectly and may also inhibit apoptosis and facilitate proliferation, invasiveness, and metastasis. Excessive production of vascular O_2 can results in hypertension and vasospasm (Lepoivre et al. 1994). Oxidative stress plays an important role in the break down immunological tolerance, inflammatory processes and can induce apoptosis and even cell death (Dalal et al. 2017; Messner and Imlay 2002; Rice-Evans and Gopinathan 1995; Dalal and Biswas 2019).

4.1.1.1 Endogenous Source

Various intracellular enzymes such as peroxisomes, lipoxygenases, NADPH oxidases, oxidases, etc. can produce the pro-oxidants inside the cells (Landry and Cotter 2014). CYP450 is a heme-containing protein superfamily that can degrade toxic compounds. Dioxygen is activated in the catalytic process CYP450 via a single electron reduction reaction to O_2 in the CYP450 catalytic process (Lewis 2002). Cytochrome P450 2E1 (CYP2E1) can generate ROS in the presence or absence of a substrate, so it is also called leaky enzyme (Robertson et al. 2001).

Lipoxygenase is a metalloenzyme used to metabolize the eicosanoid like leukotrienes and prostaglandins. Arachidonic acid is reduced into 5-Lipoxygenase (5-LO), known to be involved in the formation of leukotriene (Dixon et al. 1990). Leukotriene and 5-LO are directly related to oxidation and inflammation in arthritis, asthma, and neurodegenerative diseases (Joshi and Praticò 2015). Peroxidases are multifunctional enzymes that produce H_2O_2 which is further degraded by catalase (Nordgren and Fransen 2014; Wang et al. 2013). In mitochondria, four-electron transport chain multiprotein complexes (complex I-IV) regulate oxidative phosphorylation and the gradient of electrochemical protons (Liu et al. 2002). Complex I and III interacts with O_2 and release oxygen radical ($O_2^{\cdot}$) in the cytoplasm (St-Pierre et al. 2002).

4.1.1.2 Exogenous Source

Oxidants production can also be controlled by external environmental factors like ionization radiation, bacterial, and fungal toxins and inflammatory cytokines. Exposure of a specific cell type to these external factors may result in ROS that may affect adjacent cells also. The co-expression of multiple Toll-like receptors (TLRs) can cause oxidative stress by disrupting the generation of anti-inflammatory and pro-inflammatory cytokines (Lavieri et al. 2014). It has been reported that ionization radiation (IR) exposure to thyroid cells can generate ROS (Ameziane-El-Hassani et al. 2015). IR is one of the main causes of bonds breakage and generation of free radicals that can contribute to oxidative stress. The induction of *Streptococcus pneumonia*-mediated oxidative stress depends on LytA pneumococcal autolysin (Zahlten et al. 2014). Deoxynivalenol (DON) produced by *Fusarium* can damage the membrane and diminish the cell viability (Yang et al. 2014). DON-treated lymphocytes increase ROS levels, 8-hydroxy-2-deoxyguanosine, and lipid peroxide levels.

4.1.2 Antioxidants

Antioxidants play a major role in counteracting oxidants, which contributes to a further reduction in oxidative stress. Thus, antioxidant and oxidant balance may be used to assess human oxidative stress. Antioxidants possess antitumor, anti-carcinogenic, anti-inflammatory, antibacterial, antiviral, and antiatherosclerotic properties (Owen et al. 2000). Natural antioxidants reduce the risk of diabetes, cancer, and cardiovascular diseases and can, however, cause oxidative DNA damage. Antioxidants can be endogenous or exogenous (natural or synthetic). Both types of antioxidants have the ability to eliminate or scavenge free radicals, which is vital for the generation of ROS. Natural antioxidants can be further diversified into two groups: enzymatic and non-enzymatic, as shown in Fig. 4.1.

4.2 High Altitude Mediated Oxidative Stress

Aerobic metabolism is a necessary consequence for the production of reactive oxygen and nitrogen species (RONS). RONS plays a major role as a physiological or natural modulator of cellular redox milieu, further act as a signal for controlling factors of unknown and known pathophysiological and physiological processes. Increased aerobic metabolism or hyperoxia is generally assumed to easily produce a high level of RONS, that can result in oxidative damage of lipids, DNA, and proteins. Physical exercise over a certain duration or intensity can cause oxidative damage to various organs (Radak et al. 2001).

Nevertheless, the increment in the production level of RONS does not only seem to due to mitochondrial respiration, as the anaerobic activity could also cause oxidative damage (Radak et al. 1998). In addition, the defense of endothelium through the use of superoxide dismutase (SOD) has avoided the oxidative damage of lipids and the activity of xanthine oxidase, which indicates a variety of sources

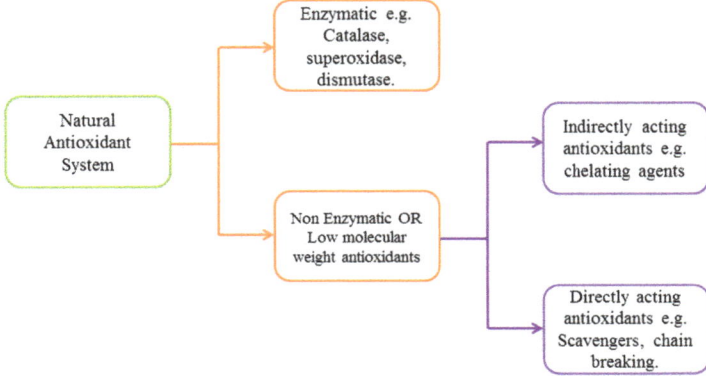

Fig. 4.1 Classification of natural antioxidants

and pathways in the creation of RONS-associated exercise. High altitude toxicity can also cause oxidative damage to macromolecules such as DNA, protein, and lipids. Low oxygen pressure appears to be beneficial for the production of less RONS, but high altitude exposure along with oxidative damage can trigger the RONS production and decrease the antioxidant system activity.

4.2.1   High Altitude and Oxidative Damage

Exposure to intermittent high altitude can significantly enhance the lipid peroxidation in fast and slow muscle fibers of rats (Radak et al. 1994). Radak et al. 1997 did not found an increment in lipid peroxidation after continuous exposure of 4 weeks, however, the amount of protein oxidation detected by carbonyl derivatives was increased (Radák et al. 1997; Kumar et al. 1999). In addition, Nakanishi and colleagues reported the increment in the level of malondialdehyde level in serum, liver, lung, kidney, and heart at an altitude of 5500 m (Nakanishi et al. 1995). It has been reported that 12 healthy subjects to 4559 m of altitude causing major increases in urine-determined DNA strand breaks (Møller et al. 2001). It has been found that simultaneous exposure at a high altitude of 2700 m and cold enhances DNA damage and lipid peroxidation (Schmidt et al. 2002). At 6000 m, the rate of lipid peroxidation rose by 23% and at 8848 m by 79% reveals that the level of oxidative stress is proportional to the rise in altitude (Joanny et al. 2001). Therefore, high altitude causes oxidative damage to proteins, DNA, and lipids by an increment in the level of generation of ROS or a decline in antioxidant capacity.

4.2.2   RONS Generation at High Altitude

The large supply of oxygen can enhance the production level of mitochondrial ROS. It also seems, however, that hypoxia also appears to lead to less stress, which can also result in an increment in the production of ROS (Mohanraj et al. 1998). The reductive stress can increase ROS production by automotive oxidation of mitochondrial complexes. It has been reported that the cellular level of NADH/NAD+ ratio increases during reductive stress (Khan and O'Brien 1995).

During hypoxia conditions, the xanthine dehydrogenase/oxidase system is a powerful ROS generator. High altitude irregular exposure has similar properties to ischemia/reperfusion (Radak et al. 1994). However, the changes in ROS and NO pattern during ischemia/reperfusion and high altitude exposure are different. The initial response is followed by a reversible increment in ROS production during ischemia/reperfusion and is reversed by antioxidants, which can raise the NO in tissue. Unlike ischemia/reperfusion, ROS level rises in hypoxia and return to pre-hypoxic values in normoxia. Acclimatization required inducible NO synthase (iNOS) regulation suggests that hypoxia can alter the ROS/NO balance (Gonzalez and Wood 2001). This phenomenon can influence the microcirculation correlated with acute mountain sickness, hypoxic exposure, high altitude brain edema, and lung. Serrano et al. reported that the presence of different NOS forms in NO formation at high altitudes might lead to a rise in the production level of nitrotyrosine in rat cerebellum (Serrano et al. 2003).

4.2.3 Hypoxia and Oxidative Stress

The exposure of high altitude can enhance the level of production of ROS and RNS, which can alter the redox balance (Magalhães et al. 2005). It has been shown that hypoxic exposure for short and long term can raise the level of oxidative stress (Joanny et al. 2001; Askew 2002; Dosek et al. 2007). Both types of hypoxia; hypobaric (i.e. terrestrial altitude) and normobaric hypoxia (i.e., simulated altitude) can increase oxidative stress (Magalhães et al. 2005; Debevec et al. 2014). However, hypobaric hypoxia found to triggers a higher level of oxidative stress than normobaric hypoxia (Faiss et al. 2013; Damij et al. 2015; Ribon et al. 2016). Three different mechanisms found to be directly involved in ROS modulation in normobaric and hypobaric hypoxia (Fagerberg 2018). First, hypobaric breathing is lower as compared to normobaric hypoxia along with higher respiratory frequency and lower tidal volume (Faiss et al. 2013; Savourey et al. 2003). In hypobaric hypoxia, higher alveolar physiological dead space is linked with hypocapnia and ventilator alkalosis. Second, higher hypoxemia due to hypobaric hypoxia may also negatively correlated between the level of oxidative stress and hemoglobin oxygen saturation (Bailey et al. 2001). Lastly, hypobaric exhaled NO levels were lower than normobaric hypoxia (Hemmingsson and Linnarsson 2009). Oxidative stress due to environmental hypoxia relies on its duration and intensity (Debevec et al. 2014; Damij et al. 2015). It seems, in general, that major deleterious effects of hypoxia-induced oxidative stress caused by hypoxia are due to high hypoxic doses. It is found that exogenous antioxidant does not appear to mitigate oxidative stress caused by hypoxia throughout exposure at high altitudes (Subudhi et al. 2004).

4.3 Rheumatoid Arthritis (RA)

Rheumatoid arthritis (RA) is a chronic autoimmune disease that proliferates the synovial cells at the joint. A high amount of infiltrates of macrophages, B cells, T cells, and polymorphonuclear cells are found at the inflammatory sites. These cells and cellular factors exhibit a major role in joint destruction in RA. RA involves environmental factors and genetic factors that can activate autoimmune responses. RA is characterized by swelling, heat, inflammation, redness, and joint pain. It causes the proliferation of synovial cells and tissues which is destructive to the cartilage and bone. The autoantibodies present in the serum of RA patients are responsible for the autoimmune reactions in the body.

4.3.1   Oxidative Stress in RA

Macrophage and polymorphonuclear cells play a vital role in the development of ROS by activation of inflammatory molecules that can destroy bones and cartilage in humans (Mapp et al. 1995). In RA, oxidative stress includes many factors such as lymphocytes, autophagy induction, and irregular production of adhesive molecules along with cell damage (Ozkan et al. 2007). Numerous antioxidants such as metallothioneins, thioredoxin, and glutathione (GSH) reductase may be present in RA synovial tissues, but due to their low level, they cannot overcome oxidative stress.

In RA patients, thioredoxin can cause synovial fibroblast cells to produce TNF α induced IL-6 and IL-8. Vitamin C, thiols, and glutathione (GSH) blood levels declined, while Malondialdehyde (MDA) levels in RA patients rose in comparison to the average individual. $O_2^{\cdot-}$, ROS, and HO^- expression levels are elevated in peripheral blood neutrophils and the synovial fluid of RA patients (Kundu et al. 2012). The ROS produced in neutrophils is directly linked to RA (Kundu et al. 2012). Immune cell invasion within RA joints will lead to the creation of RNS/ROS species that further activate the redox responsive pathways, migrate multiple irregular molecules expressed on lymphocytes in synovial fluid of RA patients. Role of oxidative stress in rheumatoid arthritis is shown in Fig. 4.2.

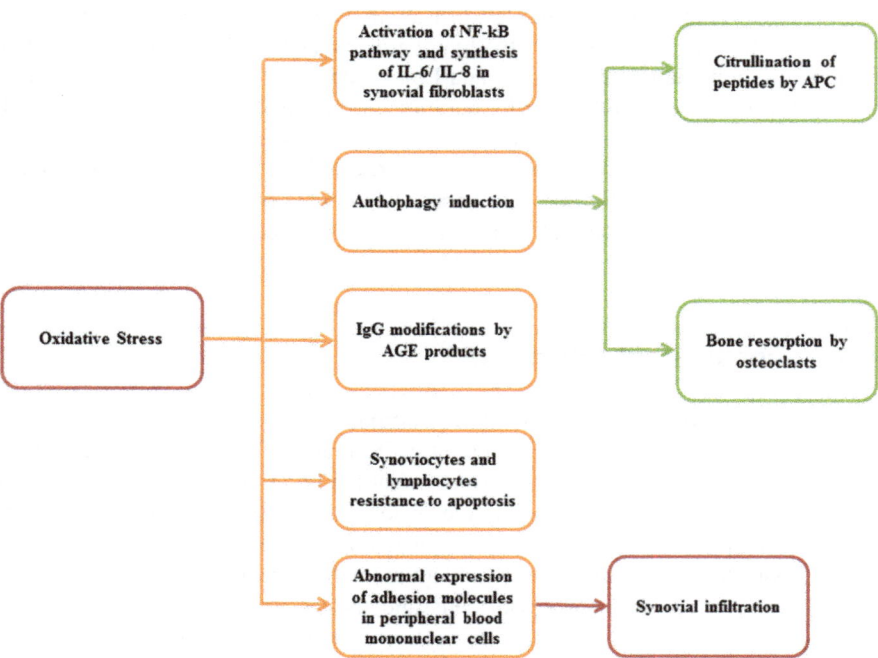

Fig. 4.2 Role of oxidative stress in rheumatoid arthritis

In RA, oxidative stress can cause the modification of immunoglobulin (IgG) (Newkirk et al. 2003). It has been found that advanced glycation end product (AGE) pentosidine and AGE-modified IgG are directly correlated with RA (Kurien and Scofield 2008). Inhibited Caspase 3 activation has not been found in the synovial cells of RA patients (Migita et al. 2001). Induction of autophagy can minimize the apoptosis in the synovial cells of RA patients (Xu et al. 2013). The levels of H_2O_2, thiobarbituric acid reactive substances (TBARS), and O_2 found higher in RA patients as compared to control healthy persons. In RA, the levels of plasma thioredoxin (TRX) and urinary excretion of 8-hydroxydeoxyguanosine (8-OHdG) were higher as compared to healthy controls (Jikimoto et al. 2002).

In RA, ROS generation is known to be directly linked to bone restoration in inflammation processes (Bijlsma and Jacobs 2000). The intracellular ionic environment may be disrupted by hypoxic conditions, which can further alter the levels of calcium and phosphorus (Cheeseman and Slater 1993). In the RA peripheral blood, the rise in lipid peroxidation can induce oxidative stress (Walwadkar et al. 2006). Moreover, lipid peroxidation can generate the MDA, which results in the generation of immunogenic molecules. In RA patients, the level of nitric oxide and lipid peroxide is higher as compared to a healthy individual. While, the concentration of calcium/phosphorus and vitamin E is lower in RA patients than healthy control. The increment in the level of nitric oxide and lipid peroxide and a decline in vitamin E and calcium/phosphorus ratio confirmed the threat of oxidative stress in rheumatoid arthritis. It has been reported that increased oxidative stress induces T cells to trigger various stimuli that can regulate immune responses and even can cause severe problems (Hassan et al. 2011). Thirty different antioxidants and oxidants were identified in RA patients. These were classified as: (1) lipid peroxidation, (2) protein oxidation, (3) DNA damage, (4) urate oxidation, (5) enzymatic activity, (6) antioxidants, and (7) free radical/anions (Quiñonez-Flores et al. 2016).

4.3.2   High Altitude and RA

It has been reported that climatic changes have a worsening effect on arthritis (Holbrook 1960). In RA, the period after the storm or rainfall is most painful (Singh et al. 1977). It has been found that a few hour exposures of rising in humidity and a decline in pressure can have a significant worsening effect on RA (Hollander and Yeostros 1963). Higher humidity and dropping barometric pressure are followed by intracellular fluid diuresis and extrusion of intracellular fluid into the blood. Diseased tissues lost its permeability, retain fluids, and therefore maintain higher intracellular pressure as compared to surrounding tissue, which further results in increased pain and swelling. Due to this, RA patients benefit from the warm and dry Southwest climate of the USA (Singh et al. 1977). It has been reported that patients at 35% humidity and 32 °C also improved (Edström 1944).

Men exposed to a dry and alternate warm and cold environment does not have adversely affect due to enhanced immune response and fibrinolytic activity rather than from meteorological variations. The excess of deposition of fibrin results in inflammation in RA, which further causes an elevation in the deposition of more fibrinogen (Fearnley et al. 1966). Exposure of high altitude can enhance the excretion of 17-hydroxysteriod in the urine which can elevate the synthesis of corticosteroids that may play a pivotal role in the prevention of RA.

4.3.3 Reactive Species Measurement in RA

Several reports show that different biomarkers like advanced oxidation of protein products (AOPP), 3-Chlorotyrosine (CT), and nitrosothiols can be used for protein oxidation evaluation in RA (Datta et al. 2014; Stamp et al. 2012; Tetik et al. 2010). It has been found that the rate of carbonylation of protein is higher in the plasma samples of RA as compared to healthy (Stamp et al. 2012; Tetik et al. 2010). In RA patients, protein carbonyls (PC), AOPP, and RNS have been determined (Datta et al. 2014). It has been reported that the level of 3-Chlorotyrosine (CT) is more in RA than a healthy person (Nzeusseu Toukap et al. 2014). Malondialdehyde (MDA) is the end product of lipid peroxidized decomposition reactions. It has been found that the MDA level is enhanced in the synovial fluid of RA patients (Gambhir et al. 1997). The fluorometric method can be utilized to measure plasma lipid peroxidation level in RA (Conti et al. 1991).

The activity of GSH-Px can be determined by spectrophotometric at 37 °C and 412 nm (Gambhir et al. 1997). The spectrophotometer can be used to measure the hydrogen peroxide and molybdate at 405 nm (Gambhir et al. 1997). The levels of antioxidants in plasma of RA can be measured by automated calorimetric methods (Erel 2004). The intracellular NO can be measured by a non-fluorescent dye diaminofluorescein diacetate (DAF-2DA), which fluorescent after reaction with NO (Sarkar et al. 2011).

4.4   Oxidative Stress Measurement

Various techniques such as HPLC, GC-MS, UV-spectroscopy, and immunoassays can be used to determine the concentration of principal biomarkers of lipid, DNA, and protein oxidative damage, as shown in Table 4.2.

The level of oxidative stress can be calculated by detecting the concentration of RS. RS can be measured Ex vivo, in vivo, or inside cells. Various techniques such as L-band electron spin along with nitroxy probe and magnetic resonance imaging spin can be used to detect RS directly inside the cells (Berliner et al. 2001). RS can be measured directly or indirectly by measuring the concentration of generated NO^- and $H_2O_2^-$ RS can be detected by measuring the concentration of trapped species or oxidative damage.

Electron spin resonance (ESR) can measure the free radicals directly by detecting the unpaired electrons. However, reactive radicals cannot be detected by ESR as they do not accumulate to a sufficient level for measurement. This problem was solved with the introduction of probes or trap agents that can form stable reactive radicals

Table 4.2 Principal markers of oxidative stress and their detection techniques

Markers	Techniques	Matrices
8-Oxo-guanine	GC-MS	DNA
	HPLC-ECD	Urine, DNA
8-oxo-2'-deoxy-guanosine	HPLC-ECD	Urine, DNA
5-(hydroxymethyl) uracil	GC	DNA, synthesized oligonucleotides
8-hydroxy-deoxy-guanosine	HPLC ECD	Urine, DNA
Hydroperoxides	Enzymatic methods	Plasma
	HPLC-MS	Plasma
	HPLC-CL	Tissue, plasma, cellular membranes
	Iodometric methods	Plasma, cellular membranes
	GC-MS	Cellular membranes
Isoprostanes	Immunoassay	Urine
	GC-MS	Plasma, tissue, urine
Malondialdehyde	HPLC	Plasma
	TBA test	Plasma, serum, tissue
4-hydroxynonenal	HPLC	Plasma, tissue
	GC-MS	Plasma, tissue, urine
Malondialdehyde	HPLC	Plasma
	TBA test	Plasma, serum, tissue
	GC-MS	Plasma, serum, tissue
Isoprostanes	Immunoassay	Urine
	GC-MS	Plasma, tissue, urine
	Radioimmunoassay	Plasma, urine
4-hydroxynonenal	HPLC	Plasma, tissue
	GC-MS	Plasma, tissue, urine

along with unstable radicals that can be detected easily by ESR. Spin traps use the hydroxylamine probes to detect the free radicals in the liver and skin (Haywood et al. 1999). It can detect the generation of secondary radicals such as lipid produced (peroxyl, alkoxyl, etc.) and protein radicals also. Aromatic traps, including phenylalanine and salicylate, are more effective than spin traps (Ingelman-Sundberg et al. 1991). Salicylate and phenylalanine can be used to assess the development of radical in RA patients (Liu et al. 1997). In Saliva, phenylalanine was used to detect the concentration of the generation of $OH^.$.

Dichlorofluorescein diacetate (DCFDA) is widely used for the detection of cellular peroxidases, although it interacts very slowly with lipid peroxidases or H_2O_2 (Ischiropoulos et al. 1999). DCFDA converts into dichlorofluorescein (DCF) which can be seen at 525 nm. Dihydrorhodamine 123 (DHR) was used to detect NO_2, $OH^.$, and $ONOO^-$, etc., although it reacts poorly to $NO.$, $O_2^-.$, and H_2O_2 (Buxser et al. 1999). DHR converted into rhodamine123 is fluorescent at 536 nm. Dihydroethidium is oxidized into a fluorescent product (ethidium) that can fluorescent at 600 nm after excitation at 500–530 nm (Fig. 4.3) (Zhao et al. 2003). Ethidium can detect the $O_2^-.$ and intercalate into nuclear DNA.

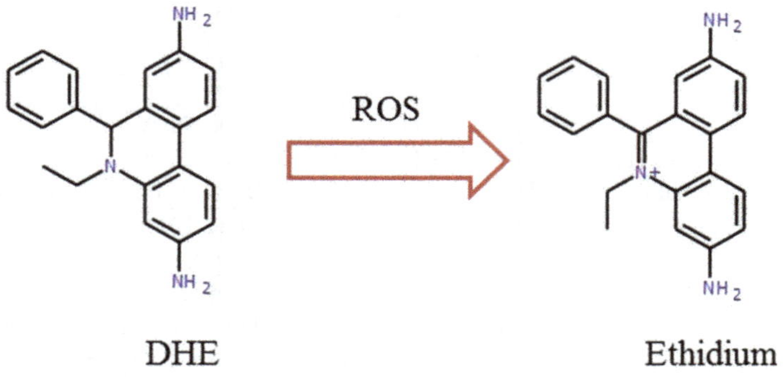

Fig. 4.3 Conversion of DHE to Ethidium

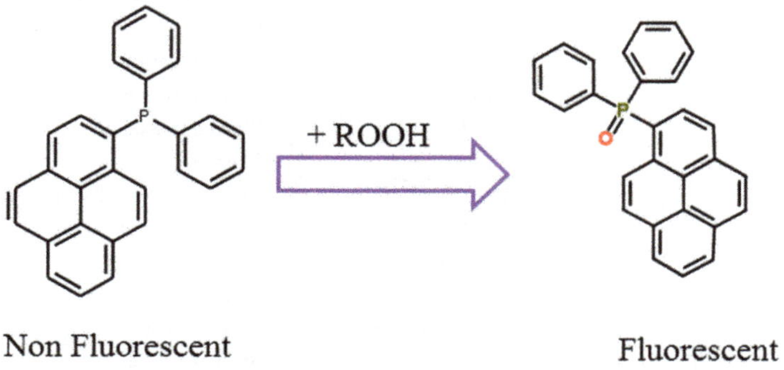

Fig. 4.4 Conversion of diphenyl-1-pyrenylphosphine (non-fluorescent) to a fluorescent product

Luminol can be used to measure the concentration of RS developed through phagocytosis activation (Faulkner and Fridovich 1993). However, luminol cannot measure O_2^- directly, it reacts with O_2^- and generate the fluorescent product. Diphenyl-1-pyrenylphosphine (DPPP) can react to peroxides and make the fluorescent product that can be detected at 380 nm after excitation at 351 nm (Fig. 4.4) (Takahashi et al. 2001). Lipid soluble hydroxide can react with Diphenyl-1-pyrenylphosphine while it cannot react with hydrogen peroxides. The fluorescent DPPP is quite stable in living cells and remained up to 2 days, whereas other minor effects such as cell morphology, proliferation, or cell viability remained up to 3 days.

Fluorescence plate reader is the simplest method that measures the variation in fluorescence. However, the machine's sensitivity and efficiency vary enormously, and the addition of extra filters and other parts will make it costly. Flow cytometry provides the benefit of measuring the cellular culture directly by fluorescence. Quantitative data can be collected on the number of cells that emits fluorescence at a certain wavelength. Although it has a drawback, the addition of trypsin can result in the development of oxidative stress. In mouse liver, the concentration of phosphatidylethanolamine (PEOOH) and phosphatidylcholine (PCOOH) was calculated by chemiluminescence-high performance liquid chromatography (CL-HPLC) (Miyazawa et al. 1987). Lipid peroxidation produce the endoperoxides, hydroperoxides, and final products: ethane, pentane, so it is the most accurate method of determination of ROS. It has been reported that determination of concentration of PEOOH or PCOOH is one of the most accurate methods of lipid peroxidation analysis (Miyazawa et al. 1987).

4.5 High Altitude and Proteomic Alteration

Exposure of hypobaric hypoxia increases the mandatory biological processes that can enhance the oxygen delivery along with cardiac output, ventilation, and hematocrit in lowlanders rising to the altitude (Peacock 1998). Similarly, physiological traits that can increase oxygen flux have been selected in high altitude populations (Beall 2007). However, there is a pattern of acclimatization, which may vary between highland populations such as Tibetans has higher resting ventilation rates as compared to Andeans, while arterial oxygen contents and hematocrits are lower than Andeans or lowlanders (Beall 2007). Exhaled vasodilator nitric oxide and signal molecules are higher in Andeans as compared to lowlanders and Tibetans (Beall et al. 2001). Several variants in the GTP-cyclohydrolase 1 gene involved in the stabilization of NO synthase have been enhanced in Tibetans with high circulating NO levels. Furthermore, NO can promote pulmonary perfusion and provide protection for pulmonary hypertension as experienced at altitude by outlanders (Busch et al. 2001). In Tibetan, elevated circulatory NO metabolites are also linked to the increased flux of limb blood flow and NO itself can lead to the modulation of hematocrit, reducing blood viscosity (Erzurum et al. 2007; Ashmore et al. 2014).

A genomic analysis in the Tibetan highlanders demonstrated a peroxisome proliferator-activated receptor (PPARA) haplotype positively selected and correlated with the phenotype of a lower hematocrit (Simonson et al. 2010). Peroxisome proliferator-activated receptor α (PPARα) encoded by PPARA, which plays an important role in the regulation of cell metabolism. PPARα is expressed in liver, heart, and muscle and can enhance the expression of fatty acid metabolism controlling genes (Gulick et al. 1994; Gilde and Van Bilsen 2003). PPARA haplotype is correlated with an increment of non-esterified fatty acids, which can result in a decline of whole-body fatty acid oxidation (FAO) in Tibetan (Ge et al. 2012). Whereas, in Sherpas, PPARA haplotype is co-related with a decrease in the expression of skeletal muscle PPARα and carnitine palmitoyltransferase 1 (CPT1B) which can result in a decrease in mitochondrial FAO capacity (Horscroft et al. 2017). In hypoxia, cellular oxygen requirements may be reduced due to switch in the substrate of ATP synthesis from fatty acid to non-fatty acids. It has been reported that reduction in FAO capacity and increment in mitochondrial coupling efficiency after some time of altitude exposure in native lowlanders make them adaptable (Horscroft et al. 2017; Jacobs et al. 2012).

The increment in glycolytic flux in highlander and lowlander populations can activate hypoxia induce factor (HIF) to induce lactate efflux and glycolysis in the cells (Semenza et al. 1994; Kim et al. 2006; Papandreou et al. 2006). Enhancement in lactate dehydrogenase (LDH) activity indicates the increment in cardiac glucose uptake and lactate efflux capacity in Sherpas as compared to lowlanders (Horscroft et al. 2017; Holden et al. 1995). Therefore, increment in glucose metabolism, especially glycolysis is a function of adaptation and acclimatization to high altitude.

4.6 Conclusion

The disturbance between the production of antioxidants and oxidants in a biological system can result in oxidative stress. Oxidative stress is one leading cause of apoptosis which can cause in cell death. Four oxidants development sources are: aerobic respiration, phagocytosis of bacteria or virus, H_2O_2 production by peroxisome, and cytochrome P450. Oxidants are mutagenic in nature and play a major role in invasiveness, metastasis or suppression of apoptosis. Arachidonic acid is reduced into 5-LO which can cause inflammation in asthma, arthritis, and neurodegenerative conditions. Specific environmental factors, such as ionizing radiations, bacterial and fungal toxins, and inflammatory cytokines, can also play an important role in the regulation of development of oxidants. Antioxidants are required to counter the oxidants which can decrease oxidative stress. Antioxidants exhibit antitumor, anti-carcinogenic, anti-inflammatory, antibacterial, and antiviral properties.

The development of reactive oxygen and nitrogen species required an aerobic metabolism process. Increment in hyperoxia or aerobic metabolism can produce a high level of RONS, which can result in oxidative stress. Physical exercise after a certain intensity or duration can cause oxidative damage to several organs. Exposure of high altitude can increase the rate of lipid peroxidation in fast and slow muscle fibers. Even short exposure at high altitudes can increase lipid peroxidation. An abundant supply of oxygen can enhance the development of mitochondrial ROS.

Rheumatoid arthritis (RA) is a chronic autoimmune condition that proliferates in the articulation of the synovial cells. Autoantibodies in the serum of RA patients are responsible for the body's autoimmune reactions. Rheumatoid factor is known as an autoantibody for RA diagnosis, however, it is present in two-third patients only. Several other antibodies, such as heterogeneous nuclear RNPs, mannose-binding lectin (MBL), and immunoglobulin binding protein (BiP) can also be used for the detection of RA. Polymorphonuclear cells and macrophages may induce ROS, the generation of ROS can trigger chronic inflammation which can destruct the human bone and cartilage. The reduction in antioxidants levels of blood increases the chance of RA development. A few hours of elevated humidity and reduction in pressure will greatly exacerbate RA effects. Oxidative stress in RA can be measured by protein oxidation or lipid oxidation by detection of various biomarkers such as AOPP, RSNO, CT, and MDA, etc.

Several techniques such as GC-MS, HPLC, and UV-spectroscopy, etc. can be used to measure the concentration of reactive species. Reactive cell species can also be detected by using different compounds such as Dihydrorhodamine 123 (DHR), Diphenyl-1-pyrenylphosphine (DPPP), or luminol, etc. Different techniques such as flow cytometry, confocal microscopy, and fluorescence microplate reader, etc. have been used to detect the oxidative by measurement of the production of reactive species. Physiological acclimatization can be observed in lowlanders at altitude. These differences between individuals are due to genetic difference among them. Hypobaric hypoxia can enhance the biological process, resulting in an increment in cardiac output, ventilation, and hematocrit. Increased altitude can alter the expression or activity of various proteins such as PPARA, LDH, HIF, and CPT1B, etc.

In the last decade, research has been done on the detection of free radicals. Several techniques, along with sensors or probes, have been identified. There is a requirement of the development of new sensors or probes to measure the ROS within a human cell. The production of molecules that can inhibit oxidants or activate antioxidants is highly required. New biomarkers are required to detect the proteomics alteration at high altitude.

References

Ameziane-El-Hassani R, Talbot M, Dos Santos MCDS, Al Ghuzlan A, Hartl D, Bidart J-M et al (2015) NADPH oxidase DUOX1 promotes long-term persistence of oxidative stress after an exposure to irradiation. Proc Natl Acad Sci 112(16):5051–5056

Ashmore T, Fernandez BO, Evans CE, Huang Y, Branco-Price C, Griffin JL et al (2014) Suppression of erythropoiesis by dietary nitrate. FASEB J 29(3):1102–1112

Askew E (2002) Work at high altitude and oxidative stress: antioxidant nutrients. Toxicology 180(2):107–119

Bailey D, Davies B, Young IS (2001) Intermittent hypoxic training: implications for lipid peroxidation induced by acute normoxic exercise in active men. Clin Sci (Lond) 101:465–475

Beall CM (2007) Two routes to functional adaptation: Tibetan and Andean high-altitude natives. Proc Natl Acad Sci 104(suppl 1):8655–8660

Beall C, Laskowski D, Strohl KP, Soria R, Villena M, Vargas E, Alarcon AM, Gonzales C, Erzurum SC (2001) Pulmonary nitric oxide in mountain dwellers. Nature 414:411–412

Berliner LJ, Khramtsov V, Fujii H, Clanton TL (2001) Unique in vivo applications of spin traps. Free Radic Biol Med 30(5):489–499

Bijlsma JW, Jacobs JW (2000) Hormonal preservation of bone in rheumatoid arthritis. Rheum Dis Clin 26(4):897–910

Busch T, Bartsch P, Pappert D, Grunig E, Hildebrandt W, Elser H et al (2001) Hypoxia decreases exhaled nitric oxide in mountaineers susceptible to high-altitude pulmonary edema. Am J Respir Crit Care Med 163(2):368–373

Buxser SE, Sawada G, Raub TJ (1999) Analytical and numerical techniques for evaluation of free radical damage in cultured cells using imaging cytometry and fluorescent indicators. In: Methods in enzymology. Elsevier, Amsterdam, pp 256–275

Cheeseman K, Slater T (1993) An introduction to free radical biochemistry. Br Med Bull 49(3): 481–493

Conti M, Morand P, Levillain P, Lemonnier A (1991) Improved fluorometric determination of malonaldehyde. Clin Chem 37(7):1273–1275

Dalal V, Biswas S (2019) Nanoparticle-mediated oxidative stress monitoring and role of nanoparticle for treatment of inflammatory diseases. In: Nanotechnology in modern animal biotechnology. Elsevier, Amsterdam, pp 97–112

Dalal V, Sharma NK, Biswas S (2017) Oxidative stress: diagnostic methods and application in medical science. In: Oxidative stress: diagnostic methods and applications in medical science. Springer, Berlin, pp 23–45

Damij N, Levnajić Z, Skrt VR, Suklan J (2015) What motivates us for work? Intricate web of factors beyond money and prestige. PLoS One 10(7):e0132641

Datta S, Kundu S, Ghosh P, De S, Ghosh A, Chatterjee M (2014) Correlation of oxidant status with oxidative tissue damage in patients with rheumatoid arthritis. Clin Rheumatol 33(11): 1557–1564

Debevec T, Pialoux V, Mekjavic IB, Eiken O, Mury P, Millet GP (2014) Moderate exercise blunts oxidative stress induced by normobaric hypoxic confinement. Med Sci Sports Exerc 46(1): 33–41

Dixon R, Diehl R, Opas E, Rands E, Vickers P, Evans J et al (1990) Requirement of a 5-lipoxygenase-activating protein for leukotriene synthesis. Nature 343(6255):282

Dosek A, Ohno H, Acs Z, Taylor AW, Radak Z (2007) High altitude and oxidative stress. Respir Physiol Neurobiol 158(2–3):128–131

Edström G (1944) Can rheumatic infection be influenced by an artificial tropical climate? Acta Med Scand 117(3–4):376–414

Erel O (2004) A novel automated method to measure total antioxidant response against potent free radical reactions. Clin Biochem 37(2):112–119

Erzurum S, Ghosh S, Janocha A, Xu W, Bauer S, Bryan N et al (2007) Higher blood flow and circulating NO products offset high-altitude hypoxia among Tibetans. Proc Natl Acad Sci 104(45):17593–17598

Fagerberg P (2018) Negative consequences of low energy availability in natural male bodybuilding: a review. Int J Sport Nutr Exerc Metab 28(4):385–402

Faiss R, Pialoux V, Sartori C, Faes C, Dériaz O, Millet GP (2013) Ventilation, oxidative stress, and nitric oxide in hypobaric versus normobaric hypoxia. Med Sci Sports Exerc 45(2):253–260

Faulkner K, Fridovich I (1993) Luminol and lucigenin as detectors for O2⊡−. Free Radic Biol Med 15(4):447–451

Fearnley G, Chakrabarti R, Evans J (1966) Fibrinolytic treatment of rheumatoid arthritis with phenformin plus ethyloestrenol. Lancet 288(7467):757–761

Gambhir JK, Lali P, Jain AK (1997) Correlation between blood antioxidant levels and lipid peroxidation in rheumatoid arthritis. Clin Biochem 30(4):351–355

Ge R-L, Simonson TS, Cooksey RC, Tanna U, Qin G, Huff CD et al (2012) Metabolic insight into mechanisms of high-altitude adaptation in Tibetans. Mol Genet Metab 106(2):244–247

Gilde A, Van Bilsen M (2003) Peroxisome proliferator-activated receptors (PPARS): regulators of gene expression in heart and skeletal muscle. Acta Physiol Scand 178(4):425–434

Gonzalez NC, Wood JG (2001) Leukocyte-endothelial interactions in environmental hypoxia. In: Hypoxia. Springer, Berlin, pp 39–60

Gulick T, Cresci S, Caira T, Moore DD, Kelly DP (1994) The peroxisome proliferator-activated receptor regulates mitochondrial fatty acid oxidative enzyme gene expression. Proc Natl Acad Sci 91(23):11012–11016

Hassan SZ, Gheita TA, Kenawy SA, Fahim AT, El-Sorougy IM, Abdou MS (2011) Oxidative stress in systemic lupus erythematosus and rheumatoid arthritis patients: relationship to disease manifestations and activity. Int J Rheum Dis 14(4):325–331

Haywood RM, Wardman P, Gault DT, Linge C (1999) Ruby laser irradiation (694 nm) of human skin biopsies: assessment by electron spin resonance spectroscopy of free radical production and oxidative stress during laser depilation. Photochem Photobiol 70(3):348–352

Hemmingsson T, Linnarsson D (2009) Lower exhaled nitric oxide in hypobaric than in normobaric acute hypoxia. Respir Physiol Neurobiol 169(1):74–77

Holbrook W (1960) Climate and the rheumatic diseases. In: Hollander JL (ed) Arthritis and allied conditions. Henry Kimpton, London, pp 577–581

Holden J, Stone C, Clark C, Brown W, Nickles R, Stanley C et al (1995) Enhanced cardiac metabolism of plasma glucose in high-altitude natives: adaptation against chronic hypoxia. J Appl Physiol 79(1):222–228

Hollander JL, Yeostros SJ (1963) The effect of simultaneous variations of humidity and barometric pressure on arthritis. Bull Am Meteorol Soc 44(8):489–494

Horscroft JA, Kotwica AO, Laner V, West JA, Hennis PJ, Levett DZ et al (2017) Metabolic basis to Sherpa altitude adaptation. Proc Natl Acad Sci 114(24):6382–6387

Ingelman-Sundberg M, Kaur H, Terelius Y, Persson J, Halliwell B (1991) Hydroxylation of salicylate by microsomal fractions and cytochrome P-450. Lack of production of 2, 3-dihydroxybenzoate unless hydroxyl radical formation is permitted. Biochem J 276(3): 753–757

Ischiropoulos H, Gow A, Thom SR, Kooy NW, Royall JA, Crow JP (1999) Detection of reactive nitrogen species using 2, 7-dichlorodihydrfluorescein and dihydrorhodamine 123. In: Methods in enzymology. Elsevier, Amsterdam, pp 367–373

Jacobs RA, Siebenmann C, Hug M, Toigo M, Meinild A-K, Lundby C (2012) Twenty-eight days at 3454-m altitude diminishes respiratory capacity but enhances efficiency in human skeletal muscle mitochondria. FASEB J 26(12):5192–5200

Jikimoto T, Nishikubo Y, Koshiba M, Kanagawa S, Morinobu S, Morinobu A et al (2002) Thioredoxin as a biomarker for oxidative stress in patients with rheumatoid arthritis. Mol Immunol 38(10):765–772

Joanny P, Steinberg J, Robach P, Richalet J, Gortan C, Gardette B et al (2001) Operation Everest III (Comex'97): the effect of simulated severe hypobaric hypoxia on lipid peroxidation and antioxidant defence systems in human blood at rest and after maximal exercise. Resuscitation 49(3):307–314

Joshi YB, Praticò D (2015) The 5-lipoxygenase pathway: oxidative and inflammatory contributions to the Alzheimer's disease phenotype. Front Cell Neurosci 8:436

Khan S, O'Brien PJ (1995) Modulating hypoxia-induced hepatocyte injury by affecting intracellular redox state. Biochim Biophys Acta Mol Cell Res 1269(2):153–161

Kim J-w, Tchernyshyov I, Semenza GL, Dang CV (2006) HIF-1-mediated expression of pyruvate dehydrogenase kinase: a metabolic switch required for cellular adaptation to hypoxia. Cell Metab 3(3):177–185

Kumar D, Bansal A, Thomas P, Sairam M, Sharma S, Mongia S et al (1999) Biochemical and immunological changes on oral glutamate feeding in male albino rats. Int J Biometeorol 42(4): 201–204

Kundu S, Ghosh P, Datta S, Ghosh A, Chattopadhyay S, Chatterjee M (2012) Oxidative stress as a potential biomarker for determining disease activity in patients with rheumatoid arthritis. Free Radic Res 46(12):1482–1489

Kurien BT, Scofield RH (2008) Autoimmunity and oxidatively modified autoantigens. Autoimmun Rev 7(7):567–573

Landry WD, Cotter TG (2014) ROS signalling, NADPH oxidases and cancer. Portland Press, London

Lavieri R, Piccioli P, Carta S, Delfino L, Castellani P, Rubartelli A (2014) TLR costimulation causes oxidative stress with unbalance of proinflammatory and anti-inflammatory cytokine production. J Immunol 192(11):5373–5381

Lepoivre M, Flaman J, bobe P, Lemaire G, Henry Y. (1994) Quenching of the tyrosil free radical of ribonucleotide reductase by nitric oxide. J Biol Chem 269:21891–21897

Lewis DFV (2002) Oxidative stress: the role of cytochromes P450 in oxygen activation. J Chem Technol Biotechnol 77(10):1095–1100

Liu L, Leech JA, Urch RB, Silverman FS (1997) In vivo salicylate hydroxylation: a potential biomarker for assessing acute ozone exposure and effects in humans. Am J Respir Crit Care Med 156(5):1405–1412

Liu Y, Fiskum G, Schubert D (2002) Generation of reactive oxygen species by the mitochondrial electron transport chain. J Neurochem 80(5):780–787

Magalhães J, Ascensão A, Soares JM, Ferreira R, Neuparth MJ, Marques F et al (2005) Acute and severe hypobaric hypoxia increases oxidative stress and impairs mitochondrial function in mouse skeletal muscle. J Appl Physiol 99(4):1247–1253

Mapp P, Grootveld M, Blake D (1995) Hypoxia, oxidative stress and rheumatoid arthritis. Br Med Bull 51(2):419–436

Messner KR, Imlay JA (2002) Mechanism of superoxide and hydrogen peroxide formation by fumarate reductase, succinate dehydrogenase, and aspartate oxidase. J Biol Chem 277(45): 42563–42571

Migita K, Yamasaki S, Kita M, Ida H, Shibatomi K, Kawakami A et al (2001) Nitric oxide protects cultured rheumatoid synovial cells from Fas-induced apoptosis by inhibiting caspase-3. Immunology 103(3):362–367

Miyazawa T, Yasuda K, Fujimoto K (1987) Chemiluminescence-high performance liquid chromatography of phosphatidylcholine hydroperoxide. Anal Lett 20(6):915–925

Mohanraj P, Merola AJ, Wright VP, Clanton TL (1998) Antioxidants protect rat diaphragmatic muscle function under hypoxic conditions. J Appl Physiol 84(6):1960–1966

Møller P, Loft S, Lundby C, Olsen NV (2001) Acute hypoxia and hypoxic exercise induce DNA strand breaks and oxidative DNA damage in humans. FASEB J 15(7):1181–1186

Nakanishi K, Tajima F, Nakamura A, Yagura S, Ookawara T, Yamashita H et al (1995) Effects of hypobaric hypoxia on antioxidant enzymes in rats. J Physiol 489(3):869–876

Newkirk MM, Goldbach-Mansky R, Lee J, Hoxworth J, McCoy A, Yarboro C et al (2003) Advanced glycation end-product (AGE)-damaged IgG and IgM autoantibodies to IgG-AGE in patients with early synovitis. Arthritis Res Ther 5(2):R82

Nordgren M, Fransen M (2014) Peroxisomal metabolism and oxidative stress. Biochimie 98:56–62

Nzeusseu Toukap A, Delporte C, Noyon C, Franck T, Rousseau A, Serteyn D et al (2014) Myeloperoxidase and its products in synovial fluid of patients with treated or untreated rheumatoid arthritis. Free Radic Res 48(4):461–465

Owen R, Giacosa A, Hull W, Haubner R, Spiegelhalder B, Bartsch H (2000) The antioxidant/anticancer potential of phenolic compounds isolated from olive oil. Eur J Cancer 36(10): 1235–1247

Ozkan Y, Yardým-Akaydýn S, Sepici A, Keskin E, Sepici V, Simsek B (2007) Oxidative status in rheumatoid arthritis. Clin Rheumatol 26(1):64–68

Papandreou I, Cairns RA, Fontana L, Lim AL, Denko NC (2006) HIF-1 mediates adaptation to hypoxia by actively downregulating mitochondrial oxygen consumption. Cell Metab 3(3): 187–197

Peacock AJ (1998) Oxygen at high altitude. BMJ 317(7165):1063–1066

Quiñonez-Flores CM, González-Chávez SA, Del Río ND, Pacheco-Tena C (2016) Oxidative stress relevance in the pathogenesis of the rheumatoid arthritis: a systematic review. Biomed Res Int 2016:6097417

Radak Z, Lee K, Choi W, Sunoo S, Kizaki T, Oh-Ishi S et al (1994) Oxidative stress induced by intermittent exposure at a simulated altitude of 4000 m decreases mitochondrial superoxide dismutase content in soleus muscle of rats. Eur J Appl Physiol Occup Physiol 69(5):392–395

Radák Z, Asano K, Lee K-C, Ohno H, Nakamura A, Nakamoto H et al (1997) High altitude training increases reactive carbonyl derivatives but not lipid peroxidation in skeletal muscle of rats. Free Radic Biol Med 22(6):1109–1114

Radak Z, Nakamura A, Nakamoto H, Asano K, Ohno H, Goto S (1998) A period of anaerobic exercise increases the accumulation of reactive carbonyl derivatives in the lungs of rats. Pflugers Arch 435(3):439–441

Radak Z, Taylor AW, Ohno H, Goto S (2001) Adaptation to exercise-induced oxidative stress: from muscle to brain. Exerc Immunol Rev 7:90–107

Ribon A, Pialoux V, Saugy J, Rupp T, Faiss R, Debevec T et al (2016) Exposure to hypobaric hypoxia results in higher oxidative stress compared to normobaric hypoxia. Respir Physiol Neurobiol 223:23–27

Rice-Evans CA, Gopinathan V (1995) Oxygen toxicity, free radicals and antioxidants in human disease: biochemical implications in atherosclerosis and the problems of premature neonates. Essays Biochem 29:39

Robertson G, Leclercq I, Farrell GC II (2001) Cytochrome P-450 enzymes and oxidative stress. Am J Physiol Gastrointest Liver Physiol 281(5):G1135–G11G9

Sarkar A, Saha P, Mandal G, Mukhopadhyay D, Roy S, Singh SK et al (2011) Monitoring of intracellular nitric oxide in leishmaniasis: its applicability in patients with visceral leishmaniasis. Cytometry A 79(1):35–45

Savourey G, Launay JC, Besnard Y, Guinet AL, Travers SP (2003) Normo-and hypobaric hypoxia: are there any physiological differences. Eur J Appl Physiol 89:122–126

Schmidt MC, Askew E, Roberts DE, Prior RL, Ensign W Jr, Hesslink RE Jr (2002) Oxidative stress in humans training in a cold, moderate altitude environment and their response to a phytochemical antioxidant supplement. Wilderness Environ Med 13(2):94–105

Semenza GL, Roth PH, Fang H-M, Wang GL (1994) Transcriptional regulation of genes encoding glycolytic enzymes by hypoxia-inducible factor 1. J Biol Chem 269(38):23757–23763

Serrano J, Encinas JM, Salas E, Fernandez AP, Castro-Blanco S, Fernández-Vizarra P et al (2003) Hypobaric hypoxia modifies constitutive nitric oxide synthase activity and protein nitration in the rat cerebellum. Brain Res 976(1):109–119

Simonson T, Yang Y, Huff C, Yun H, Qin G, Witherspoon D, Jorde LB, Prchal JT, Ge R (2010) Genetic evidence for high-altitude adaptation in Tibet. Science 329:72–660

Singh I, Chohan I, Lal M, Khanna P, Srivastava M, Nanda R et al (1977) Effects of high altitude stay on the incidence of common diseases in man. Int J Biometeorol 21(2):93–122

Stamp LK, Khalilova I, Tarr JM, Senthilmohan R, Turner R, Haigh RC et al (2012) Myeloperoxidase and oxidative stress in rheumatoid arthritis. Rheumatology 51(10):1796–1803

St-Pierre J, Buckingham JA, Roebuck SJ, Brand MD (2002) Topology of superoxide production from different sites in the mitochondrial electron transport chain. J Biol Chem 277(47): 44784–44790

Subudhi AW, Jacobs KA, Hagobian TA, Fattor JA, Fulco CS, Muza SR et al (2004) Antioxidant supplementation does not attenuate oxidative stress at high altitude. Aviat Space Environ Med 75(10):881–888

Takahashi M, Shibata M, Niki E (2001) Estimation of lipid peroxidation of live cells using a fluorescent probe, diphenyl-1-pyrenylphosphine. Free Radic Biol Med 31(2):164–174

Tetik S, Ahmad S, Alturfan AA, Fresko I, Disbudak M, Sahin Y et al (2010) Determination of oxidant stress in plasma of rheumatoid arthritis and primary osteoarthritis patients. Indian J Biochem Biophys 47:353–358

Walwadkar S, Suryakar A, Katkam R, Kumbar K, Ankush R (2006) Oxidative stress and calcium-phosphorus levels in rheumatoid arthritis. Indian J Clin Biochem 21(2):134

Wang B, Van Veldhoven PP, Brees C, Rubio N, Nordgren M, Apanasets O et al (2013) Mitochondria are targets for peroxisome-derived oxidative stress in cultured mammalian cells. Free Radic Biol Med 65:882–894

Xu K, Xu P, Yao J-F, Zhang Y-G, Hou W-k, Lu S-M (2013) Reduced apoptosis correlates with enhanced autophagy in synovial tissues of rheumatoid arthritis. Inflamm Res 62(2):229–237

Yang W, Yu M, Fu J, Bao W, Wang D, Hao L et al (2014) Deoxynivalenol induced oxidative stress and genotoxicity in human peripheral blood lymphocytes. Food Chem Toxicol 64:383–396

Zahlten J, Kim Y-J, Doehn J-M, Pribyl T, Hocke AC, García P et al (2014) Streptococcus pneumoniae–induced oxidative stress in lung epithelial cells depends on pneumococcal autolysis and is reversible by resveratrol. J Infect Dis 211(11):1822–1830

Zhao H, Kalivendi S, Zhang H, Joseph J, Nithipatikom K, Vásquez-Vivar J et al (2003) Superoxide reacts with hydroethidine but forms a fluorescent product that is distinctly different from ethidium: potential implications in intracellular fluorescence detection of superoxide. Free Radic Biol Med 34(11):1359–1368

Oxidative Stress, ROS Generation, and Associated Molecular Alterations in High Altitude Hypoxia

5

Aditya Arya and Shikha Jain

Abstract

High altitude especially above 3000 m of elevation is considered a potential risk for rapid ascent, while at elevation above 5500 m, it poses life threatening risk to non-adapted individuals. This threat is associated with hypoxia-induced rapid cardiopulmonary changes that impair the normal functioning of the body. It is well-evident that partial pressure of oxygen is much reduced at altitudes due to thinning of the atmosphere and therefore, the human body prefers to adapt against those depriving oxygen levels. In this attempt, some unbalanced changes in subcellular physiology may lead to alterations and hence cause reversible and irreversible damage. Among the most rapid cellular perturbations during altitude-associated hypoxia is the generation of reactive oxygen species (ROS). Although cells have mechanisms to succumb to those altered levels of ROS, the inability to quickly alleviate ROS leads to enormous damage to proteins, lipids, and other biomolecules. As cellular response cells readjust their redox milieu and strengthen antioxidant defence. The dietary intervention of antioxidants has also proven to be useful in clinical settings. This chapter describes the mechanism of ROS generation and ROS associated proteomic perturbation at the cellular level during hypobaric hypoxia.

Keywords

Reactive oxygen species · High altitude · Hypobaric hypoxia · Antioxidants

A. Arya (✉)
National Institute of Malaria Research, Indian Council of Medical Research, New Delhi, India

S. Jain
Department of Oral Biology, School of Dental Medicine, University at Buffalo, Buffalo, NY, USA

© The Author(s), under exclusive license to Springer Nature Singapore Pte Ltd. 2022
N. K. Sharma, A. Arya (eds.), *High Altitude Sickness – Solutions from Genomics, Proteomics and Antioxidant Interventions*,
https://doi.org/10.1007/978-981-19-1008-1_5

5.1   Introduction

Environment plays a pivotal role in determining the well-being of organisms. The quality of life and longevity depends on the quality of the surrounding environment. Several environmental stressors are known to impact the quality of life and can lead to life-threatening diseases. Environmental stressors may be classified as natural or man-made. Man-made environmental stressors are primary pollutants that vary in severity and concentration depending on human activity. Some of the major man-made environmental stressors include high carbon monoxide in the air, pesticides in soil, and heavy metals in drinking water. These stressors may be avoided and their ill effects may be prevented by purification at the source. However, natural stressors are difficult to evade and they need suitable precautions in the form of individual adaptation or prophylactic modalities. Natural environmental stressors may arise from reduced air pressure at mountains (hypobaric hypoxia), increased pressure in deep-sea diving, cold temperature, high temperatures, UV radiations, etc. Figure 5.1 illustrates the major environmental stressors.

Among the natural environmental stressors, most of them are caused due to varying geological architecture of earth, primarily altitudes, deep oceans, or even latitudinal variations. Varying latitudes across the equator shows a gradation in temperature creating a huge variation of +55 °C maximum temperature in some

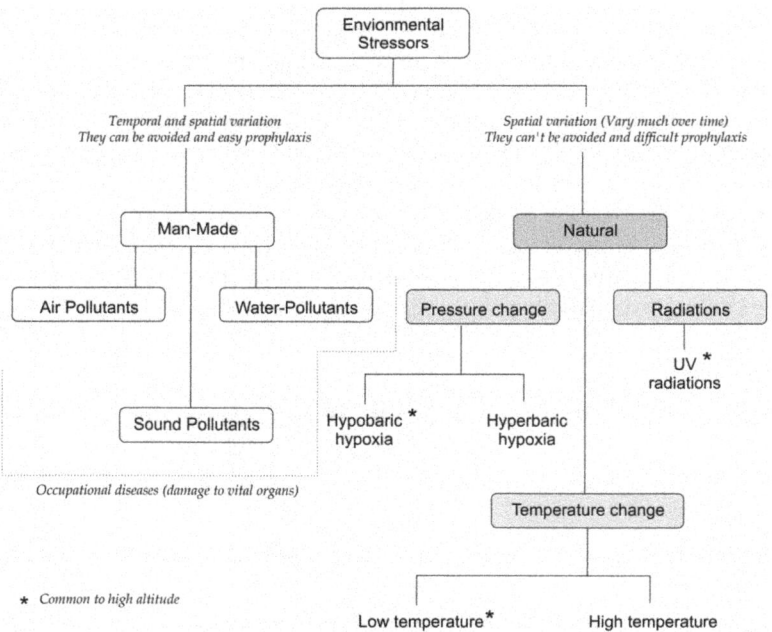

Fig. 5.1 Major types of environmental stressors: man-made stressors can be evaded while natural stressors are difficult to evade and need prophylactic or therapeutic interventions

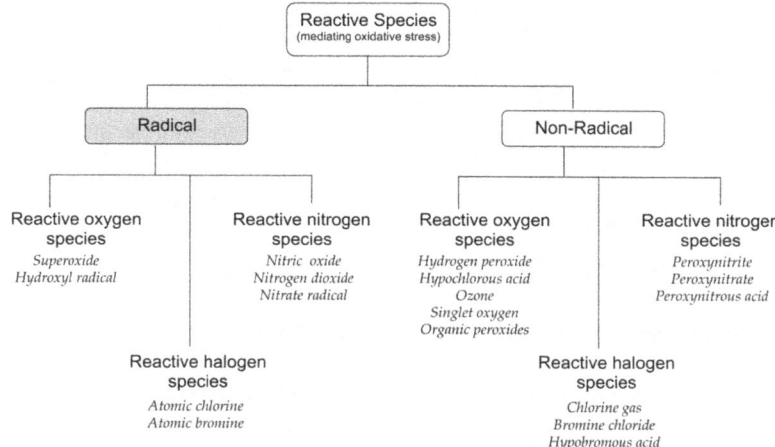

Fig. 5.2 Common Pro-oxidants explored in biological systems

countries of Africa while lowest temperatures of -50 °C at poles. High pressure exists as a potent stressor in deep oceans. Another factor is altitude, which poses two parallel threats, one in the form of reduced air pressure and another as elevated UV radiations, creating it one of the most difficult terrains for organism survival. These changes, above a threshold, induce the compensatory adaptive cardiopulmonary responses, primarily hyperventilation and generation of excess radicals as a result of an imbalance of glucose metabolism and oxygen availability resulting in the onset of high altitude sickness.

Chemically, various molecules such as oxygen have the ability to accept electrons are potentially called as oxidant or oxidising agents (Prior and Cao 1999) and this phenomenon of electron removal is called as oxidation. Although the process of oxidation is concurrent with reduction in another paired molecule or group, hence a more general term redox reaction is used. Redox reactions build the foundations of key biochemical pathways including biosynthesis and energy production. They are also important in understanding biological oxidation and radical/antioxidant effects. While the terms oxidant and reductant are primarily of chemical origin, in biological environments pro-oxidant and antioxidant terms are often preferred (Kohen and Nyska 2002). Pro-oxidants include both radical and non-radical species (Halliwell 2006). A few most abundant and common pro-oxidants are listed in Fig. 5.2.

A number of aforementioned environmental stressors are known to directly or indirectly influence the conventional redox milieu and impair the steady state of oxidant and antioxidants, this condition is referred to as oxidative stress (Halliwell 2006; Kalyanaraman 2013). Moreover, recently, the oxidative stress has been further elaborated to distinguish the thin line of difference between oxidative imbalance that causes pathological conditions called distress, while the one that constitutes an essential part of redox signalling is called oxidative eustress. A number of pathological conditions especially those influenced by environmental stressors are known to have involvement of oxidative stress and are therefore characterised by the presence of hallmarks of oxidative stress or redox-biomarkers (Dalle-Donne et al. 2006). Among the two most common environmental stressors known to cause oxidative stress are hypoxemia and UV radiation. The former disrupts the oxygen flux in mitochondria and affects the steady flow of electrons via electron transport chain and hence promotes reactive oxygen species generation. While in case of UV radiations, the high energy photos are directly known to induce homolytic cleavage of bonds and therefore generation of radicals across various biomolecules.

5.2   Mechanisms of Oxidant Generation

Living organisms are continuously at the risk of being exposed to reactive oxidants of both extrinsic and intrinsic origin, some of which are related with the random interaction of molecular oxygen and its trade-off (Winterbourn 2008). Generation of RONS in living cells can be attributed to either direct emergence from interaction of high energy radiations (such as UV radiations at high altitude) or via mitochondrial route due to the leakage of electrons from electron transport chain (Novo and Parola 2008; Lin and Beal 2006). A detailed discussion on mitochondrial ROS generation system is provided in the following description.

5.2.1   Oxidants Generation in Mitochondria

Nearly, 1% to 5% of electrons flowing at a steady rate in electron transport chain could be diverted to the generation of superoxide radicals ($O_2^{\cdot-}$), which mostly occurs across the NADH/ubiquinone oxidoreductase or ubiquinol/cytochrome c oxidoreductase, otherwise commonly known as complex I and complex III of electron transport chain (Kohen and Nyska 2002). Superoxide radicals are then usually detoxified to hydrogen peroxide by mitochondrial superoxide dismutase enzymes. This hydrogen peroxide can now cross the mitochondrial membrane and reach cytoplasm, which could damage other biomolecules unless detoxified via several peroxidases (Cadenas and Davies 2000). Moreover, besides membrane bound complexes which contribute to the oxidant generation, several matrix proteins are also actively involved in oxidant generation such as alpha ketoglutarate dehydrogenase (KGDH), aminocycloproane-carboxylic acid oxidase (ACO), and pyruvate dehydrogenase complex (PDC), which involve the oxidation of NADH are primary matrix sources of radicals. Besides these common ROS-generating pathways which lead to the formation of either superoxides or hydrogen peroxides, several antioxidants counter the effect. These enzymatic or non-enzymatic antioxidant systems either independently (catalase, Mn-SOD, glutathione peroxidase) or operate as a cascade. Among a very common cascade is the oxidation–reduction cycling of Peroxiredoxins (PRX3), Thioredoxins (TRX2), Glutathione peroxidase (GPX), and Glutathione (GSH) as shown in Fig. 5.3.

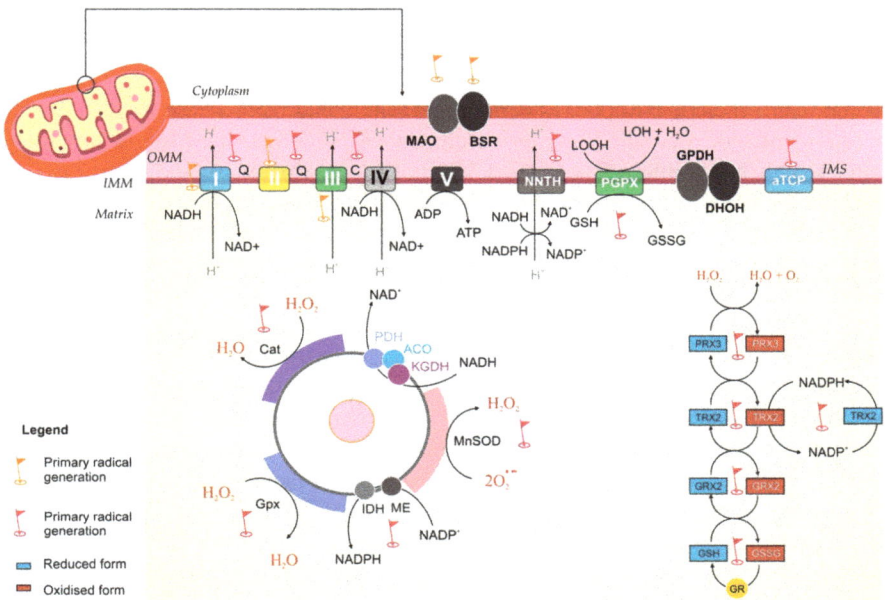

Fig. 5.3 Summary of major ROS generation systems and antioxidant defence inside mitochondria. (*OMM* outer mitochondrial membrane, *IMM* intermembrane space, *MAO* Monoamine oxidase, *PRX* peroxiredoxin, *TRX* Thioredoxin, *GPX* Glutathione peroxidase, *GSH* Glutathione (reduced), *GSSG* Glutathione (oxidised), *IDH* Isocitrate dehydrogenase, *Mn-SOD* Manganese superoxide dismutase

Although the formation of a primary ROS, i.e., superoxides occurs in the respiratory chain from molecular oxygen. This is believed to follow first order kinetics with oxygen concentration. As a paradox, however, generating ROS in mitochondria in the cell remains constant or known to be elevated with a concurrent decrease in PO_2 which is often observed with hypoxic conditions, either normobaric or hypobaric. Interestingly, the reason behind aforementioned paradox is higher affinity of molecular oxygen to ROS-generating modules compared to conventional acceptor cytochrome oxidase or coenzyme Q. However, the generation of ROS predominates during oxygen limiting condition while the cytochrome oxidase is partially reduced. It is also noteworthy to mention that this paradox is absent in the isolated mitochondrial system.

5.2.2   Oxidant Generation in Phagocytic and Non-phagocytic Cells

Besides conventional mitochondrial route, yet another important radical generation route is through the catalysis of NADPH oxidase (NOX). Enzyme NOX is present in several phagocytic cells such as macrophages, neutrophils, and eosinophils, as well as non-phagocytic cells and its presence is associated with several diseases (Babior 1999; Vignais 2002; Lambeth 2007) particularly, chronic liver diseases (CLDs)

(De Minicis and Brenner 2007). There are two membrane bound components in classical NOX of phagocytic origin namely, p22phox and gp91phox/Nox2, each of which comprises flavocytochrome and four cytosolic components. The cytosolic components include p40 phox, p47phox, p67phox, and the GTPase Rac1/2. Stimulation of phagocytic cells leads to recruitment of NOX into the plasma membrane after which NOX interacts with Cyt b558. This interaction is known to increase activity and subsequent generation of reactive oxygen species.

On the other hand, NADPH oxidase of non-phagocytic cells is similar in structure and function as that of phagocytic NOX. However, gp91phox/Nox2 is substituted by another member of the same family, usually Nox1, Nox3, Nox4, Nox5, or Duox1/2, which are homologues of Nox2. In contrast, non-phagocytic NOX is different, as it results in relatively low levels of ROS. It is observed that presence of 5-Lipoxygenase (5-LOX), a mixed function oxidase, elevates the activity and ROS generation mediated by NOX. 5-LOX is known to catalyse the synthesis of leukotrienes from arachidonic acid after some exogenous stimulus and can therefore stimulate NOX (Novo and Parola 2008). The growth factors and cytokines lead to membrane ruffling and the generation of superoxide, leading to H_2O_2, through the intervention of the small GTPase Rac1 and a SOD isoform (Soberman 2003). ROS can also be generated enzymatically in many subcellular compartments by several oxidases, peroxidases, mono- and di-oxygenases and by isoforms of the cytochrome P450 superfamily. Here it seems relevant to mention nitric oxide synthase and xanthine oxidase (Vasquez-Vivar and Kalyanaraman 2000; Pritsos 2000), cyclooxygenase (COX), and other NAD(P)H dependent oxidoreductase are also able to generate superoxide radicals. Moreover, oxidases of peroxisomal origins such as D-amino oxidases, ureate oxidases, glycolate oxidases, and fatty acid-CoA oxidases can generate hydrogen peroxide (H_2O_2) during their routine metabolic reactions (Rojkind et al. 2002). Also, an enzyme lysyl oxidase that catalyses the formation of the aldehyde precursors forming cross-links in collagen and elastin can also give rise to H_2O_2 as a result of electron leakage (Fig. 5.4).

5.3   Defence Mechanism Against Oxidative Stress

Defence against oxidative stress or rapid surge in reactive oxygen and nitrogen species is effectively countered by antioxidants. It is evident from the evolutionary process, life witnessed and adapted according to the changing oxygen concentration, in fact mostly an increase and life adapted to emerge from reduced environment to the oxidised one, suggesting the origin and prevalence of strong intrinsic antioxidant defence in most organisms. The antioxidant defence system of living organisms contains two major arms or operations, enzymatic antioxidants, and non-enzymatic antioxidants.

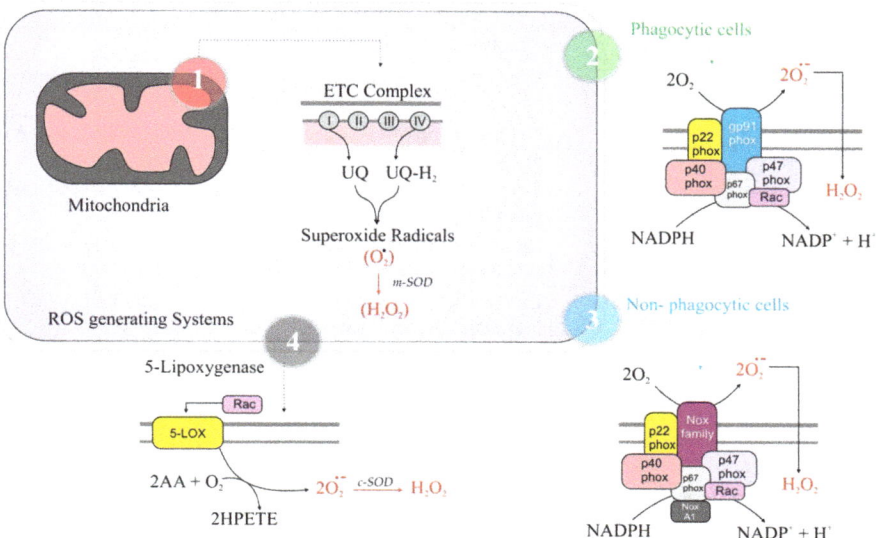

Fig. 5.4 Major endogenous sources of cellular oxidants in phagocytic cells, non-phagocytic cells, and peroxisomes

5.3.1 Enzymatic Antioxidants

A number of enzymes present in the human body are able to confer antioxidant abilities by catalysing the reduction process. One of the first lines of antioxidant defences in the cell include the activity of superoxide dismutase (SOD) which catalyses the conversion of superoxide into less reactive peroxides. SOD is compartmentalised as two different isoforms, namely Mn-SOD (SOD I), that is localised in mitochondria, second Cu-Zn-SOD (SOD II), localised in cytoplasm and EC-SOD (SOD III) localised in extracellular spaces. Among the most common dismutation of superoxides is the formation of hydrogen peroxide (H_2O_2) or sometimes organic peroxides. These peroxides are then detoxified with the help of catalase, and peroxidases in various compartments. Catalase is the key enzyme that catalyses this neutralisation of hydrogen peroxide, perhaps due to its high K_m value which allows it to remain active even at high concentrations of H_2O_2. In contrast to catalase, another enzyme peroxidase has relatively lower K_m and therefore remains active at low H_2O_2 concentrations. Peroxidases, depending on the requirement of glutathione are further grouped as glutathione (GSH) dependent peroxidase and glutathione independent peroxidase, the latter class includes, thioredoxin (Trx) dependent called peroxiredoxin (Prx). Peroxiredoxins are particularly important for antioxidant defence in erythrocytes (Rhee et al. 2005). Some of the commonly occurring cellular antioxidant enzymes, their activities and associated enzyme commission numbers are summarised in Table 5.1.

Table 5.1 Description of common enzymatic antioxidant defence operating during hypoxia

Enzyme	Reaction catalysed
Superoxide dismutase (EC 1.15.1.1)	Conversion of superoxide radicals in less toxic hydrogen peroxide
Catalase (EC 1.11.1.6)	Neutralisation of hydrogen peroxide in water (with a high K_m or low affinity)
Glutathione peroxidase (1.11.1.12)	Detoxification of organic peroxides (especially PUFA-OOH) into fatty acid and water
Glutathione S-transferases (2EC .5.1.18)	Reduction of proteins and other active molecules by addition of reducing groups of glutathione
Phospholipid-hydroperoxide glutathione peroxidase (EC 1.11.1.9)	Detoxification of organic peroxides (especially PUFA-OOH) into fatty acid and water
Ascorbate peroxidase (EC 1.11.1.11)	Neutralisation of hydrogen peroxide in water with concurrent oxidation of ascorbate into dihydroascorbate
Gualacol type peroxidase (EC 1.11.1.7)	Neutralisation of organic peroxides, halides or sulphates into less toxic forms
Monodehydroascorbate reductase (EC 1.6.5.4)	NADH mediated and recycling of dihydroascorbate (oxidised ascorbate) to ascorbic acid
Dehydroascorbate reductase (EC 1.8.5.1)	Glutathione mediated and recycling of dihydroascorbate (oxidised ascorbate) to ascorbic acid
Glutathione reductase (EC 1.6.4.2)	NADPH mediated recycling of oxidised glutathione (GSSG) into reduced glutathione (GSH)

5.3.2 Non-Enzymatic Antioxidants

Glutathione and thioredoxin are the most abundant among the non-enzymatic antioxidants. Glutathione is a dipeptide having cysteine in its structure that augments glutathione peroxidase activity and maintains cellular proteins in their reduced state. This process is expensive for the cells as NADPH is consumed during the conversion of oxidised glutathione (GSSG) to reduced glutathione (GSH) by an enzyme glutathione reductase (GR). Thioredoxin is another protein that augments peroxiredoxin activity and maintains protein thiolation. The intracellular picture of radical generation and its scavenging mechanisms is illustrated in Fig. 5.4.

Apart from these non-enzymatic antioxidants, especially metabolites such as melatonin and uric acid and dietary organic molecules such as vitamin E, lycopene, vitamin C, carotenes, and several polyphenols have shown potentials to scavenge the radicals and therefore alleviate oxidative stress above the basal threshold (Seifried et al. 2007; Catoni et al. 2008). On the basis of the above discussion, the entire antioxidant defence system may be stratified into four layers, the first is the immediate defence at the source that directly scavenges ROS, the next could be the first line of defence that depends on antioxidant enzymes. Then, the second line of defence including ancillary factors that augment conventional antioxidant defence and finally, the third line of defence including small metabolites and dietary antioxidants such as metal chelators and vitamins (Fig. 5.5).

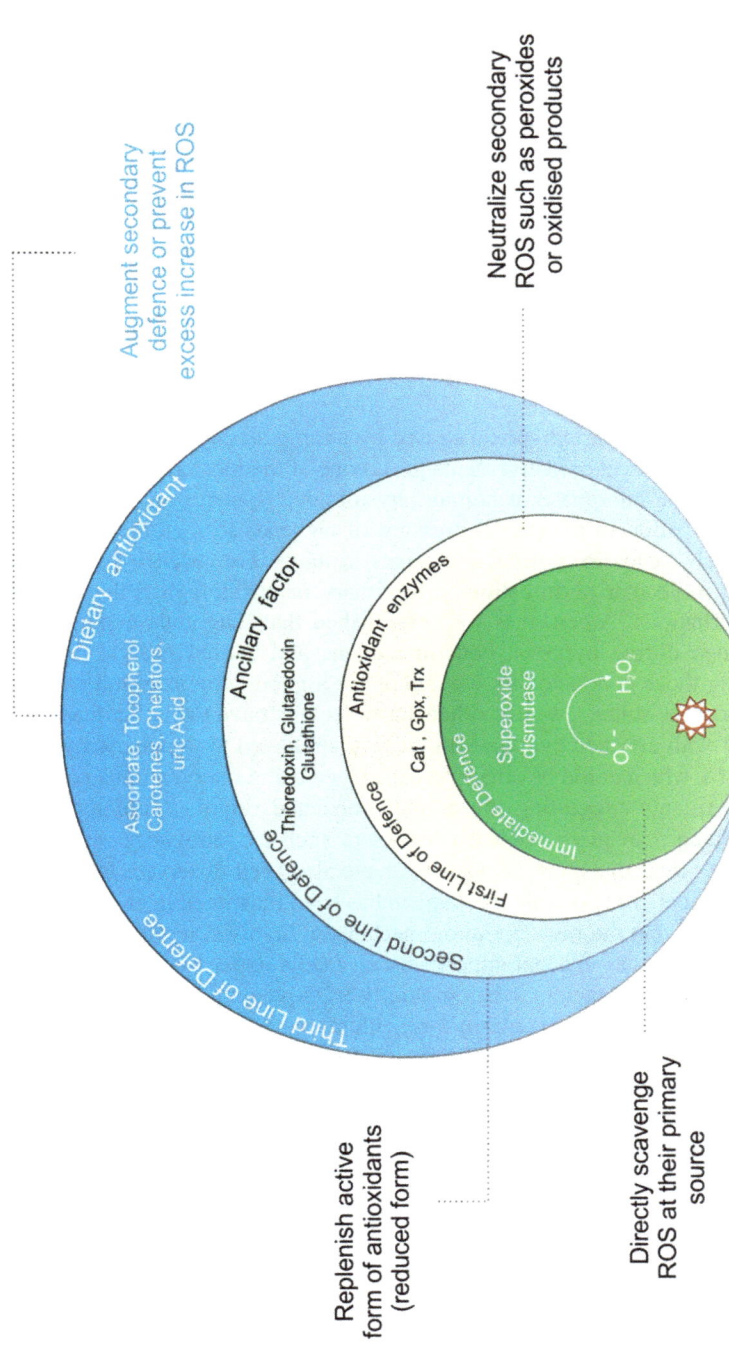

Fig. 5.5 Defence against oxidative stress can be broadly stratified into three levels, primary, secondary, and tertiary. The first level directly dimutates highly evanescent ROS such as superoxides by the presence of enzymes like superoxide dismutase, while the next level operates at neutralising peroxides, either hydrogen peroxide or organic peroxides by the action of peroxiredoxins, glutathione peroxidases, or catalase. The third strata of antioxidant defence mostly relies on dietary antioxidants including uric acid ascorbate, tocopherol, or carotenes

5.4   Organ-Level Oxidative Damage Manifested by Cellular ROS

Although the generation of ROS and RNS is manifested by alterations in the intracellular redox centres, the effects are manifested at organ levels. There are several reports of tissue level and organ level damage due to oxidative stress caused by hypobaric hypoxia. Among the most vulnerable organs are the lung and brain. The brain contains a large proportion of lipids that are at the direct risk of modification by oxidants, while lungs which stay at the interphase of atmosphere and body remain vulnerable due to change in external pressure.

5.4.1   Oxidative Damage in Lungs

Hypoxia, especially hypobaric hypoxia caused by a rapid ascent to altitudes is also associated with organ-level oxidative damage and one of the most affected organs is lung. Lung lies at the interphase of human physiological systems and outer atmosphere, any perturbation in the partial pressure of air leads to a cardiopulmonary burden and thereby leads to molecular changes in lungs. However, it is not well understood that molecular perturbations are primary or physiological that lead to molecular perturbations. Albeit, it is well established that lung cells experience a significant change during hypoxia, both in humans and animal models. Several studies including those conducted by the author are a good demonstration of these findings. In one of the studies, we and others have demonstrated the inflammation of lung concurrent with elevated cytokines in rat lung simulated to hypobaric hypoxia (Arya et al. 2016), which could be effectively managed using interventional of novel classes of antioxidants suggesting the oxidant mediated origin of inflammation. Furthermore, changes in physiological parameters such as ventilatory responses, cardiopulmonary blood transport mechanisms have also been observed. In a recent review by Siques et al., the possible mechanism has been illustrated in greater detail. It has been reported that the most common mechanism involved in the relationship of reactive oxygen species and pulmonary vascular cells under hypobaric hypoxic conditions include several factors, where not only ROS play a role but the complex interplay of calcium within sarcoplasmic reticulum, serotonin, endothelin-1, and interleukins. Moreover, some of the crucial transcription factors or protein interactors such as signal transducer and activator of transcription (STAT3), hypoxia-inducible factor 1 (Hif1a), and enhancer-binding protein modulate the transcriptional regulation of hypoxia-induced perturbations (Siques et al. 2018).

5.4.2   Oxidative Damage in Brain

Hypoxia, especially hypobaric hypoxia that is known to be prominent at an altitude above 5500 m, has been proven to be an external stressor affecting brain functioning. The severe decline in the blood oxygen due to lowered partial pressure of oxygen is the leading cause of affected brain functions. A well-known effect on brain functioning is the loss of memory and loss of cognition, which has been documented in several studies. We and others have observed in simulated hypoxic conditions, the loss of cognition and learning abilities in experimental rats, using Morris water maze experiments and found these changes to be concurrent with elevated reactive oxygen species in the brain (Arya et al. 2016). Moreover, these findings related to high altitude hypoxia have also been corroborated with simple hypoxia in conjugation with exercise. Devebec et al. showed that increased oxidative stress caused molecular damage and disruption of redox signalling and was found to be linked with numerous pathophysiological processes and known to exacerbate chronic diseases. Prolonged systemic hypoxia that was generally elevated by exposure to terrestrial altitude or a reduction in ambient O_2 availability was found to enhance oxidative stress and thereby modifying redox balance in healthy individuals (Debevec et al. 2017). Nevertheless, a reversal of the hypoxic condition, i.e. by utilising hyperbaric conditions, the changes might be reversed. Such experimental outcomes have recently been demonstrated that hyperbaric conditions can induce neuroplasticity and improve cognitive functions in patients suffering from anoxic brain damage (Hadanny et al. 2015).

5.4.3 Oxidative Damage in Heart

The heart is the next most vulnerable organ to hypobaric hypoxia after brain and lung. Although the changes in pulmonary ventilation patterns directly influence heart functioning both at the physiological and cellular levels. Most of the damage to the heart is due to hypobaric hypoxia. Christopher Jon Boos et al. recently provided an excellent context of hypobaric hypoxia-induced perturbations in the heart in comparison with normobaric and other physiological states. Their study showed that genuine high altitude (GHA), normobaric hypoxia (NH), and hypobaric hypoxia (HH) generate similar adaptations in the heart following acute exposure despite the reduced levels of SpO_2 with GHA and HH as compared with NH (Boos et al. 2016). The reactive oxygen species are pivotal in the development of several cardiac abnormalities and functioning. ROS levels up to a certain extent also play a role in intracellular signalling and homeostasis, but, elevated ROS, particularly after an exogenous or chemical stimulus has deleterious effects and the effects are exacerbated when the rise in ROS is not compensated by the endogenous antioxidant defence system. Several studies indicate that alteration in reactive oxygen species and therefore closely associated with inflammation and progression of cardiac infarction or hypertrophy or both. This has been confirmed in several animal models post-hypoxic exposure simulating to an altitude of 3600–7620 m (or 13–8% O_2). Among the common redox-mediated signalling pathways in the heart is hypoxia-inducible factor 1 a (HIF-1α) mediated changes, which are known to lead to hypertrophic stimuli via mitogen-activated kinase superfamily. Among a few well-established redox-regulated kinases and transcription factors are JNK, p38, and ERK which control the apoptosis and inflammation in cardiac cells.

Furthermore, recent studies suggest that redox imbalance and therefore oxidative stress can activate nuclear factor-kappa B (NF-kB) in cardiac tissue and lead to cardiac inflammation and in some cases complete cardiac failure (Pena et al. 2020).

5.5 Oxidative Perturbations Mediated Changes in Proteome

Organ-level proteomic studies have been carried out by us and other researchers on subjects exposed to hypobaric hypoxia either in natural high altitude or simulated hypobaric hypoxia. Most of these studies indicate a significant change in the global proteome profile of the lung, brain, heart, and muscles. Changes in proteomic profile have been discussed in another chapter in more detail, here we would limit ourselves to perturbations in proteome ROS in terms of their modifications. Proteins, in their native conformations, have side chains of various types that can be exposed to extrinsic factors and become vulnerable to wandering reactive oxygen species and thus show reactivity with them and get modified. Some of the common redox modifications of proteins include nitrosylation, nitration, carbonylation, sulfhydration, etc. In different types of modifications, different residues are involved, most often tyrosine, cysteine, and serine. An elaborate discussion effect of potential post-translational modification (PTMs) have been recently studied by the author and others. We reported the cross-talk between the protein nitrosylation and carbonylation in hypoxic cell culture systems and later observed several PTMs in human subjects. Authors investigated the direct and indirect interactions between nitrosylation and carbonylation especially involving two protein networks associated with coagulation and inflammation pathways, found to be interlinked with redox signalling, suggesting the role of redox PTMs in hypoxia signalling favouring tolerance and survival (Gangwar et al. 2021).

5.6 Conclusion

The involvement of redox biology in high altitude physiology and pathological conditions is highly complex and a lot more is remaining to be explored, especially the downstream effectors of ROS involving several regulatory proteins, membrane lipids, and hormones. A deeper insight of these aspects in future is likely to enlighten the accomplishment of clinical success in quick identification of hypoxic susceptibility testing and management of post-hypoxic pathological conditions.

References

Arya A, Gangwar A, Singh SK et al (2016) Cerium oxide nanoparticles promote neurogenesis and abrogate hypoxia-induced memory impairment through AMPK-PKC-CBP signaling cascade. Int J Nanomed 11:1159–1173
Babior BM (1999) NADPH oxidase: an update. Blood 93(5):1464–1476

Boos CJ, O'Hara JP, Mellor A, Hodkinson PD, Tsakirides C, Reeve N et al (2016) A four-way comparison of cardiac function with Normobaric Normoxia, Normobaric hypoxia, hypobaric hypoxia and genuine high altitude. PLoS One 11(4):1

Cadenas E, Davies KJ (2000) Mitochondrial free radical generation, oxidative stress, and aging. Free Radic Biol Med 29(3–4):222–230

Catoni C, Peters A, Schaeffer HM (2008) Life history trade-offs are influenced by the diversity, availability and interactions of dietary antioxidants. Anim Behav 76:12

Dalle-Donne I, Rossi R, Colombo R, Giustarini D, Milzani A (2006) Biomarkers of oxidative damage in human disease. Clin Chem 52(4):601–623

De Minicis S, Brenner DA (2007) NOX in liver fibrosis. Arch Biochem Biophys 462(2):266–272

Debevec T, Millet GP, Pialoux V (2017) Hypoxia-induced oxidative stress modulation with physical activity. Front Physiol 8:84

Gangwar A, Paul S, Arya A, Ahmad Y, Bhargava K (2021) Altitude acclimatization via hypoxia-mediated oxidative eustress involves interplay of protein nitrosylation and carbonylation: a redoxomics perspective. Life Sci 296:120021. https://doi.org/10.1016/j.lfs.2021.120021

Hadanny A, Golan H, Fishlev G, Bechor Y, Volkov O, Suzin G, Ben-Jacob E, Efrati S (2015) Hyperbaric oxygen can induce neuroplasticity and improve cognitive functions of patients suffering from anoxic brain damage. Restor Neurol Neurosci 33(4):471–486

Halliwell B (2006) Reactive species and antioxidants. Redox biology is a fundamental theme of aerobic life. Plant Physiol 141(2):312–322

Kalyanaraman B (2013) Teaching the basics of redox biology to medical and graduate students: oxidants, antioxidants and disease mechanisms. Redox Biol 1(1):244–257

Kohen R, Nyska A (2002) Oxidation of biological systems: oxidative stress phenomena, antioxidants, redox reactions, and methods for their quantification. Toxicol Pathol 30(6): 620–650

Lambeth JD (2007) Nox enzymes, ROS, and chronic disease: an example of antagonistic pleiotropy. Free Radic Biol Med 43(3):332–347

Lin MT, Beal MF (2006) Mitochondrial dysfunction and oxidative stress in neurodegenerative diseases. Nature 443(7113):787–795. https://doi.org/10.1038/nature05292

Novo E, Parola M (2008) Redox mechanisms in hepatic chronic wound healing and fibrogenesis. Fibrogenesis Tissue Repair 1(1):5. https://doi.org/10.1186/1755-1536-1-5

Pena E, Brito J, El Alam S, Siques P (2020) Oxidative stress, kinase activity and inflammatory implications in right ventricular hypertrophy and heart failure under hypobaric hypoxia. Int J Mol Sci 21:6421

Prior RL, Cao G (1999) In vivo total antioxidant capacity: comparison of different analytical methods. Free Radic Biol Med 27(11–12):1173–1181

Pritsos CA (2000) Cellular distribution, metabolism and regulation of the xanthine oxidoreductase enzyme system. Chem Biol Interact 129(1–2):195–208

Rhee SG, Chae HZ, Kim K (2005) Peroxiredoxins: a historical overview and speculative preview of novel mechanisms and emerging concepts in cell signaling. Free Radic Biol Med 38(12): 1543–1552

Rojkind M, Dominguez-Rosales JA, Nieto N, Greenwel P (2002) Role of hydrogen peroxide and oxidative stress in healing responses. Cell Mol Life Sci 59(11):1872–1891

Seifried HE, Anderson DE, Fisher EI, Milner JA (2007) A review of the interaction among dietary antioxidants and reactive oxygen species. J Nutr Biochem 18(9):567–579. https://doi.org/10. 1016/j.jnutbio.2006.10.007

Siques P, Brito J, Pena E (2018) Reactive oxygen species and pulmonary vasculature during hypobaric hypoxia. Front Physiol 9:865

Soberman RJ (2003) The expanding network of redox signaling: new observations, complexities, and perspectives. J Clin Invest 111(5):571–574

Vasquez-Vivar J, Kalyanaraman B (2000) Generation of superoxide from nitric oxide synthase. FEBS Lett 481(3):305–306

Vignais PV (2002) The superoxide-generating NADPH oxidase: structural aspects and activation mechanism. Cell Mol Life Sci 59(9):1428–1459

Winterbourn CC (2008) Reconciling the chemistry and biology of reactive oxygen species. Nat Chem Biol 4(5):278–286

High Altitude Induced Thrombosis: Challenges and Recent Advancements in Pathogenesis and Management

6

Tarun Tyagi and Kanika Jain

Abstract

Venous thrombosis and pulmonary embolism together form a serious disorder of accelerated and unwanted intravascular blood clot formation which is termed as venous thrombo-embolism (VTE), and can be life threatening. The exposure to high altitude hypoxic environment forms one of the lesser known risk factors for VTE. A number of human and animal studies have provided some mechanistic insights into pathogenesis of high altitude induced thrombo-embolism (HATE). Increasing evidences suggest that the molecular pathogenesis of high altitude induced thrombosis/VTE is distinct from thrombosis occurring at plains. The molecular pathogenesis and clinical management of HATE remains challenging, however, recent advancements provides some insights which are discussed with an attempt to understand molecular pathogenesis and available treatment options for this disorder.

Keywords

High altitude physiology · Thrombosis · Altitude induced thrombosis · Thrombo-embolism

T. Tyagi (✉) · K. Jain
Department of Internal Medicine, Yale School of Medicine, New Haven, CT, USA
e-mail: tarun.tyagi@yale.edu

© The Author(s), under exclusive license to Springer Nature Singapore Pte Ltd. 2022
N. K. Sharma, A. Arya (eds.), *High Altitude Sickness – Solutions from Genomics, Proteomics and Antioxidant Interventions*,
https://doi.org/10.1007/978-981-19-1008-1_6

6.1   Introduction

The pathological formation of blood clot or *thrombosis* is a serious physiological disorder; if not treated immediately, can be life threatening. Although, recognized for centuries, yet the spectrum of treatment remains highly limited and depends mainly on anticoagulant therapy which bears a significant risk of bleeding. Depending on the vessel type, thrombotic disorders can be either arterial or venous in nature, distinguishable predominantly by the sequence of events leading to the thrombosis. In brief, an arterial clot is primarily triggered by the rupture of an artherosclerotic plaque, and is called "white thrombus." On the other hand, thrombi that occur in the veins are rich in fibrin and thus called "red thrombus," largely occurring without damage to the vessel wall. Venous thrombosis (VT) occurs due to changes in the composition of the blood, changes that reduce or abolish blood flow, and/or changes to the endothelium, i.e. the Virchow's Triad. In addition, the genetic and environmental factors can increase the risk of developing venous thrombosis. Although the mechanistic detail of arterial thrombosis has been widely studied (Jackson 2011; Lippi et al. 2011), the more extensive studies are needed for the in-depth understanding of the pathogenesis of venous thrombosis (Reitsma et al. 2012; Lopez and Chen 2009).

Venous thrombosis (VT) which leads to pulmonary embolism (the condition in which blood clot from elsewhere gets stuck in lung vasculature) is known by the name of **Venous Thrombo-embolism or VTE**. VTE has an incidence of approximately 1 per 1000 in adult populations annually (Heit et al. 2016; Heit 2002, 2015), and is more dangerous than VT alone. According to an estimate, 30% of patients diagnosed with VTE die within 30 days (Heit et al. 2016; Heit 2015). The major consequences of venous thrombosis are death, recurrence, post-thrombotic syndrome, and major bleeding due to anticoagulation (Cushman 2007).

6.2   Genetic Risk Factors for VTE

Several Genetic defects leading to the increased propensity of thrombosis, sometimes referred to as *"thrombophillic disorders"* or *"thrombophillia,"* have been identified; the major ones are listed in the following.

6.2.1   Factor V Leiden

Coagulation factor V is an important protein for both coagulation and anticoagulant pathways (Dahlback 2016; Lam and Moosavi 2021) . Activated Factor V protein (factor Va) of coagulation pathway acts as a cofactor for factor Xa in prothrombinase complex which converts prothrombin to thrombin. Its inactive form functions as cofactor for Activated Protein C (APC), which mediates anticoagulant pathway by inactivating factor VIIIa. A point mutation in APC cleavage site in Factor V increases the risk of venous thrombosis; individuals heterozygous for this mutation bear a fivefold increased risk for venous thrombosis while this risk becomes 50-fold in homozygous state (Rosendaal 1999; Emmerich et al. 2001).

6.2.2   Deficiency of Protein C, Protein S and Antithrombin

APC is Vitamin K dependent protein that inhibits coagulation pathway by inactivating Factor VIII a and factor Va, both of which are important for coagulation to proceed (Esmon 1993). Since activated form of Protein C has anticoagulant activity and is important for fibrinolytic pathway, its deficiency can also cause thrombophillic state (De Stefano and Leone 1995). Protein S, another vitamin K dependent protein, functions as a cofactor to APC and is present in free or bound form in plasma; its deficiency also interrupts the activity of APC and hence favors prothrombotic state (Joshi and Jaiswal 2010; Pintao et al. 2013). Antithrombin (AT, previously known as AT III) is a multifunctional protein which inhibits essentially all enzymes of coagulation pathway (Rau et al. 2007).

6.2.3   Prothrombin Mutation

The mutation in the gene for Prothrombin or factor II (Varga and Moll 2004), another of the coagulation factor genes, does not change the structure of the molecule but elevates the plasma levels of prothrombin increasing the risk of thrombosis for lifetime (Bank et al. 2004; Poort et al. 1996). Nearly 6% of all patients of venous thrombosis bears mutation in prothrombin gene (Poort et al. 1996).

6.3   Acquired Risk Factors for VTE

6.3.1   Surgery/Trauma

The risk of thrombosis varies with the type of surgery and trauma (Agnelli 2004). Hip or knee replacement, fracture (hip or leg), and spinal cord injury are classified as strong risk factors for the development of VTE (Bergqvist et al. 1983; Anderson and Spencer 2003). Geerts et al., 1994 showed in a study that 47% of trauma patients have been found to develop DVT (Geerts et al. 1994).

6.3.2   Advanced Age

Numerous reports have associated advancing age with somewhat increased risk of thrombosis (Rosendaal 1999; Oger 2000). Although the precise cause for such a strong association is not clear, factors like reduced activity, aging of veins and thus vein valves, associated with old age are thought to contribute to the increased risk.

6.3.3 Cancer

Malignancy has been linked to thrombotic events since 1865, when Trousseau observed the tendency of cancer patients to develop thrombosis (Merli and Weitz 2017). Since then, several studies have confirmed the higher rate of venous thrombosis in cancer patients (Hisada et al. 2015; Mahajan et al. 2019; Leiva et al. 2020; Khorana et al. 2007; Timp et al. 2013).

6.3.4 Oral Contraceptives and Hormone Replacement Therapy

Daily dose of a combination of an estrogen and progesterone is the most commonly prescribed form of oral contraceptives. The absolute risk of venous thrombosis in oral contraceptive users is 2 to 3 per 10,000 per year against less than 1 in nonuser women of reproductive age (Jick et al. 2000; Vandenbroucke et al. 1994; Dragoman et al. 2018).

6.3.5 Immobilization

The link between stasis/immobilization and VTE dates back to World War II when Simpson noted pulmonary embolism in people sitting on deck-chairs for prolonged periods, taking refuge in air-raid shelters and also demonstrated that sitting posture confers a greater risk than other positions (Simpson 1940). An early autopsy study found that 15% of patients who were at bed rest for less than a week had venous thrombosis; the rate was 80% with bedrest for longer duration (Gibbs 1957; Pottier et al. 2009). DVT has been shown to occur more frequently in the paralyzed limb of a hemiplegic, rather than the unaffected limb (Warlow et al. 1976)). A recent report establishes the link between immobilization and thrombosis in which prolonged sitting (more than 12 h) in front of computer is found to cause thrombosis which is termed e-thrombosis (Beasley et al. 2003; Sueta et al. 2021). However, while it is known that immobilization among patients does increase the risk of VTE, the specific role of underlying conditions cannot be excluded from consideration.

6.4 Modern Lifestyle Risk Factors for VTE

There are some other acquired risk factors for VTE which are associated with modern lifestyle (Crous-Bou et al. 2016). These include obesity, high circulating lipids, smoking, diabetes mellitus, etc. In addition to these, acute thrombotic episodes have been allegedly associated with some trivial events like sneezing and coughing attacks, sexual intercourse, strenuous physical exercise, migraine, etc.; however, the role of such events as a true trigger in any case of thrombosis is questionable (Lippi et al. 2009).

6.4.1 Virchow's Triad

Rudolph Virchow in the year 1859, first described the occurrence of thrombosis in vein. He concluded that the three factors (popular as Virchow's Triad)—(1) blood flow, (2) vessel wall, and (3) blood composition, determine the tendency of an individual to develop thrombus. "Virchow's triad" has proved to be "seminal" for modern research on thrombosis. According to the current understanding of VTE, the rise in prothrombotic factors (including IIa, tissue factor, VIIa, VIIIa, Va, Xa, platelets) defects in anticoagulant pathways, hyperactivation/dysfunction of endothelium can all stimulate thrombus formation; which may be due to single/multiple, known/unknown, genetic or/and acquired factors.

Apart from the established risk factors, the exposure to high altitude has been associated with increased incidences of thrombotic disorders, of which the venous type are more prevalent than arterial ones (Anand et al. 2001; Dilly 2021; Whayne Jr. 2014; Gupta and Ashraf 2012). Although, known for decades but as compared to other risk factors, the understanding about the pathogenesis of high altitude induced thrombosis remains highly limited.

6.5 Hypoxia at High Altitude

The barometric pressure in the atmosphere (760 mm of mercury) varies with the height or altitude from sea level. The fall in barometric pressure at altitudes lead to a proportional decrease in the partial pressure of oxygen (PaO_2), present in atmosphere, which determines the availability of oxygen for breathing. For reference, at the Everest base camp (5300 m altitude), the PaO_2 in atmosphere becomes half of that at sea level. This decreased availability of oxygen for breathing leads to oxygen limiting conditions in body, generally referred to as hypoxia. High altitude physiology may be divided into the study of short-term changes that occur with exposure to hypobaric hypoxia (the acute response to hypoxia) and studies of longer-term acclimatization and adaptation. Hypoxic challenge, such as that posed by high altitude exposure, is countered by natural physiological mechanisms of adaptation. The major adaptive changes that occur in the body can be by either increasing the oxygen delivery to tissues (by elevating hemoglobin levels) or reducing oxygen demand. The latter mechanism appears to be preferred by body as the metabolic rate and mitochondrial reactions (that solely depends on oxygen) are observed to be reduced with fall in available oxygen (Hochachka et al. 1996). However, these adaptive mechanisms cannot completely overcome the hypoxic challenge, but do increase the chances of longer survival under hypoxia.

6.6   High Altitude Thrombo-Embolism (HATE)

The term *high altitude* refers to the terrestrial elevations over 1500 m (about 5000 ft). At such elevated altitudes, diminished oxygen partial pressure, decreased temperature, lower humidity and increased UV radiations, dehydration, etc. may result in several complications/maladies. The commonly encountered ones are acute mountain sickness, HAPE (High Altitude Pulmonary Edema) and HACE (High Altitude Cerebral Edema), which form the major high altitude illness (Basnyat and Murdoch 2003). Besides these, higher incidence of thrombotic episodes is reported at high altitude, most of which occurs in veins (Anand et al. 2001; Dilly 2021) and can thus be termed as HATE (High Altitude Thrombo-Embolism). Multiple factors including hypoxic environment, immobilization, dehydration, reduced atmospheric pressure have been suggested to be responsible for HATE by many investigators, although there is no conclusive/concrete evidence to confirm the contribution of these different stressors.

6.7   Pathogenesis of HATE

Several reports have highlighted an increased risk of venous thrombosis following exposure to high altitude environment, although there mechanism behind the pathogenesis of such events is not yet completely elucidated. Apart from low PaO_2 due to reduced atmospheric pressure (hypobaric hypoxia), other factors such as low temperature, reduced mobility (especially for troops at mountains and during air travel), and dehydration also affect the physiological systems in the body and thus may contribute to increased risk of thrombosis at HA.

For decades, studies to evaluate the role of HA in thrombotic episodes and its effect on coagulation system are being done in various settings. These settings can be broadly classified as:

1. Ascent to an elevated region,
2. Long duration air travel,
3. Simulated high altitude.

6.7.1   Ascent to an Elevated Region

Several studies to ascertain the risk of thrombosis due to stay at HA have been conducted till date. Singh and Chohan had observed a tendency of hypercoagulation in Indian troops on immediate arrival at 3600 m with the increase in the platelet count, factor X, factor XII, thrombotest activity and thrombin clotting time, along with a significant increase in plasma fibrinogen and fibrinolytic activity (Singh and Chohan 1972a, b). In 1975, Ward reported deep vein thrombosis followed by pulmonary embolism in mountain climbers and a significant rise in platelet adhesiveness was observed in patients who developed ischemic strokes at HA (Sharma

1980, 1986; Sharma et al. 1977). During a study at Andes (at 3600 m and above) involving an ascending group of 28 young men, a significant increase in platelet count and hematocrit was noted 48 h after their visit as compared to the values in them at low altitude of 600 m (Hudson 1999). There have been conflicting reports about the changes in platelet count upon high altitude induction: increase (Singh and Chohan 1972b; Sharma 1980; Kotwal et al. 2007); decrease (Chatterji et al. 1982; Vij 2009); and no difference (Sharma 1986; Maher et al. 1976; Le Roux et al. 1992). During stay above 6400 m, changes in coagulation with the increase in D-dimer levels have been described, which was suggestively attributed to endothelial cell damage (Le Roux et al. 1992).

The interest in such studies was further fueled by several reports in the recent years. Bartsch et al. observed no increase in fibrin or thrombin levels in resting mountaineers who had ascended to 4559 m on foot (Bartsch et al. 2001). However, another group reported a significant rise in fibrinogen levels along with the rise in platelet activation factors in volunteers after long-term stay above 3500 m and concluded that prolonged stay at HA leads to a hypercoagulable state (Kotwal et al. 2007). Anand et al. in 2001 reported thrombosis as a complication of long-term stay at HA with a 30% higher risk (Anand et al. 2001). In addition to this, stroke cases with rate of 13.7/1000 hospital admissions were reported from HA area as compared to 1.05/1000 from plains (Jha et al. 2002).

A big question arises in HATE pathogenesis that if the hypercoagulable state and platelet activation are also associated with VTE at plains, then whether the HA associated VTE is any different in pathogenesis. In a recently published human cross sectional study, Prabhakar et al. studied the cohort of patients who developed VTE at HA, and compared this with patient cohort who developed VTE at plains (Prabhakar et al. 2019). The changes in several blood markers including coagulation, platelet activation, inflammatory response, and others were analyzed to establish HA associated pathophysiological factors in VTE. Platelet counts were found significantly elevated in HA patients (>35% patients vs. <10% of patients at plains). Besides markers of coagulation/thrombosis such as vWF and D-dimer, they also observed elevated levels of soluble platelet activation markers (P-sel, CD40L, PF4) and inflammatory markers—IL1b, IL6, and IL10 HA patients. The hypoxia associated markers—HIF-1α and the chaperoneHSP-70 were also elevated in HA cohort. The study thus highlights that the involvement of platelet activation, endothelial activation, and inflammatory molecules becomes more pronounced in HA associated VTE as compared to VTE at plains. The role of hypoxia response pathway genes was also emphasized in genome-wide expression analysis of HA-VTE patients (Jha et al. 2018).

6.7.2 Long Haul Air Travel

Numerous studies including cohort and randomized controlled trials have been performed to evaluate the risk of thrombotic complications due to long duration (>6 h) air travel. The risk of VTE associated with air travel has been found to be more than that with "weak" risk factors (like bed rest for more than 3 days, laparoscopic surgery, obesity, increasing age) and similar to "moderate" risk factors (e.g. congestive heart failure, hormone replacement therapy, deficiency of protein C and protein S) (Philbrick et al. 2007). Philbrick and coauthors observed some interesting trends about air travel related VTE reports by reviewing 24 publications from 1999 to 2005 (Philbrick et al. 2007). Case control studies, cohort studies, and randomized controlled trials (RCTs), providing information about incidence of VTE after travel, were collectively analyzed, and further define long haul air travel as a critical risk factor for development of venous thrombosis.

In a crossover study of 71 volunteers, markers of coagulation activation and fibrinolysis (thrombin-antithrombin or TAT complex, prothrombin fragment 1 + 2 or F1 + 2, D-dimers) were measured before, during, and after 8 h flight (Schreijer et al. 2006). The study also included some volunteers with already present risk factors, i.e. factor V leiden and oral contraceptive use. Activation of coagulation and fibrinolysis was evident in some of the volunteers (especially those with respective risk factors) after flight as compared to only immobilized state and daily life. This study supported the role of hypobaric hypoxia in activating coagulation system during air travel (Schreijer et al. 2006).

Concurrently, Manucci noted an increase in thrombin formation and fibrinolysis in volunteers taken to 1200 m and then to 5060 m, by air (Mannucci et al. 2002). Air travel has been proposed to amplify the risk of venous thrombosis in individuals with an already present risk factor. The supporting evidence comes from a study on 210 patients who were found to have 16% higher risk after air travel as well as 14% higher risk in women who were previously taking oral contraceptives (Martinelli et al. 2003).

6.7.3 Simulated Hypobaric Hypoxia

In an elegant single-blind, crossover study performed in a hypobaric chamber, the effect of an 8-h seated exposure to hypobaric hypoxia on hemostasis was studied on 73 healthy volunteers (Toff et al. 2006). Individuals were exposed alternately to hypobaric hypoxia, like the conditions of reduced cabin pressure during commercial air travel (equivalent to atmospheric pressure at an altitude of 2438 m), and normobaric normoxia (control condition; equivalent to atmospheric conditions at ground level). Blood was drawn before and after exposure to assess activation of hemostasis and markers of coagulation activation, fibrinolysis, platelet activation, and endothelial cell activation were evaluated. Changes were observed in some hemostatic markers during the normobaric exposure, attributed to prolonged sitting and circadian variation (Toff et al. 2006). However, there were no significant difference in any marker after hypobaric hypoxia as compared with the normobaric normoxia exposure.

Another group studied changes coagulation of arm and leg veins by measuring various parameters and by thromboelastography pre- and post-10 h exposure to simulated normobaric hypoxia (equivalent to 2400 m) (Schobersberger et al. 2007).;

however no significant activation of coagulation system was observed. In a study by Crosby et al. 2003, eight healthy volunteers were exposed to simulated hypoxic conditions for 8 h at differing altitudes to unacclimatized (3600 m) and fully acclimatized volunteers. Again, no changes were observed in markers of coagulation activation like F1 + 2, TAT, factor VIIa following hypoxic exposure. Contrary to these reports, Bendz (2000) in his study on 12 volunteers, reported activation of coagulation after sudden exposure to simulated HA (2400 m) for 2 h (Bendz et al. 2000, 2001). This study, however, lacked an adequate control group and faced immediate criticism by another group (Bartsch et al. 2001).

A case report of a 19-year-old female who developed sinus vein thrombosis after high altitude training in a hypoxic chamber (Torgovicky et al. 2005), supports the view that HA conditions predisposes an individual to thrombotic events regardless of age as a predominant risk factor. The patient in the study was working as a training instructor in a high altitude simulation chamber 2 months prior to the event and participated in six training sessions. During training she complained of dyspnea and within few days after last hypoxic exposure she developed severe frontal headaches. The CT and MRI procedures confirmed sinus vein thrombosis which was later controlled by anticoagulant drugs (Torgovicky et al. 2005).

6.8   Evidence from Pre-Clinical Studies

Studies on humans are highly limited at the interventional front, primarily due to the limited ability to control the associated variables, and animal studies are thus required to test the hypotheses generated by human cohort studies.

In a first of its kind study, Tyagi et al. found that the simulated HA (cold surrounding plus hypoxia) exposure of 6 h was able to induce prothrombotic phenotype in rats which was marked by increased platelet adhesion and aggregation and reduced bleeding time (Tyagi et al. 2014). Observed in absence of any induced vascular injury, the results emphasize the role of systemic activation of platelets in hypoxic environment. Hypoxia, but not cold climate, was found to be primarily responsible for increased platelet activation. The platelet proteome was also found to be distinctly altered in the exposed animals and the platelet calpain regulatory subunit was shown to regulate thrombogenesis through platelet activation. Besides calpain, other prothrombotic factors such as Tissue factor and fibrinogen were also upregulated in HA exposed platelets in vivo (Tyagi et al. 2014). Interestingly, the most striking platelet aggregation changes were observed in response to ADP and not thrombin, suggesting the increased sensitivity of platelets to ADP under hypoxic conditions in vivo. This finding was also supported by an expedition study wherein platelet aggregation in 22 human subjects was measured before and after ascent to HA (5300 m); platelet aggregation was observed to be higher in response to only ADP at HA as compared to the levels at plains in the same individual (Rocke et al. 2018).

Another study demonstrated, using the rat model of VTE, that exposure to simulated HA hypoxia led to NLRP3 inflammasome induction through HIF-1α pathway (Gupta et al. 2017). The pharmacological targeting of HIF-1α through si-RNA approach was shown to limit VTE in animals. Hypoxia induced signaling was also involved in VTE pathogenesis in non (hypoxia) exposed animals, suggesting the role of stasis induced local hypoxic microenvironment in VTE pathogenesis (Gupta et al. 2017). The HIF-1α has also been implicated in thrombus recanalization in mice model of DVT (Evans et al. 2010) Moreover, platelets are proposed to activate NLRP3 inflammasome in innate immune cells including neutrophils and monocytes, thereby increasing plasma IL-1β (Rolfes et al. 2020). Importantly, it has been shown that platelets themselves can assemble NLRP3 inflammasome in sepsis and can produce and release IL-1β.

Recently, in hypoxia exposed mice, the platelet depletion prevented hypoxia driven increase in pro-inflammatory cytokines CXCL4 and CCL5 in lungs (Delaney et al. 2021).This study further strengthens the central role of platelet activation which can influence inflammatory milieu in hypoxic conditions, leading eventually to immunothrombosis. The concept of immunothrombosis has recently gained interest in the view of COVID-19 pathology, which also involves hypoxemia, platelet activation, and hyperinflammatory status. VTE episodes have been shown to occur predominantly in COVID-19 patients (Di Minno et al. 2020; Middeldorp et al. 2020). In COVID-19, the acute respiratory distress syndrome (ARDS) and/or thrombosis form major cause of patient mortality (Middeldorp et al. 2020; Althaus et al. 2021; Magro et al. 2020; Ackermann et al. 2020). The platelet-neutrophil aggregates are believed to be involved in ARDS and thus role of hypoxemia induced platelet driven inflammation cannot be ruled out (Althaus et al. 2021; Viecca et al. 2020; Martinod and Deppermann 2021; Middleton et al. 2020; Le Joncour et al. 2020). The platelets interact by P-selectin present on platelet surface which binds with PSGL-1 on neutrophils and platelet-neutrophil interaction has been demonstrated to trigger NETosis. The activated platelets have upregulated surface P-selectin and thus can cause increased formation of platelet-neutrophil aggregates. Platelet activation with increased platelet-leukocyte aggregates and altered platelet transcriptome has been demonstrated in critically ill COVID-19 patients (Manne et al. 2020).

6.9   Clinical Management of HATE

As the HA has not reached a well-recognized risk factor status for thromboembolism globally, the clinical guidelines are still not clear and detailed. The thrombotic risk patients, traveling via long haul flights are generally advised to not take any additional medication unless any symptoms are present. However, as evident from prior studies, HA travel and HA stay for months or even days can pose a greater risk than a long-haul flight (Gupta and Ashraf 2012). This warrants formation of a standard clinical guideline for HATE prophylaxis and/or treatment, separate from long haul travel. Before travel to HA, a clinical assessment of thrombotic risk is advisable. A history of thrombotic episodes or presence of one or more VTE risk factors, any planned provoking activity which involves strenuous

Table 6.1 The Well's scoring system for Deep Vein Thrombosis

Clinical feature	Assigned score
Malignancy (active cancer or palliative treatment within last 6 month)	1
Paralysis, paresis or plaster immobilized lower limb	1
Recent bedridden state for 3 or more days or surgery with 12 weeks requiring anesthesia	1
Localized tenderness along the deep venous system	1
Swelling of entire leg	1
Calf swelling (minimum 3 cm larger than that on the asymptomatic side)	1
Pitting edema of the symptomatic leg	1
Collateral non varicose superficial veins	1
Documented history of deep vein thrombosis	1
Alternative diagnosis as likely as or more than deep vein thrombosis	−2

(Adapted from Wells et al. 2003)
Total Score of 2 or more = Deep vein thrombosis likely
Total Score of less than 2 = Deep vein thrombosis less likely

Table 6.2 The Well's scoring system for Pulmonary Embolism

Clinical feature	Assigned score
Clinical signs and symptoms of DVT (leg swelling, pain, palpation of leg veins)	3
An alternative diagnosis less likely than pulmonary embolism	3
Heart rate of more than 100 beats/min	1.5
More than 3 days of immobilization or surgery in previous 4 weeks	1.5
History of DVT or pulmonary embolism	1.5
Hemoptysis	1
Malignancy (within last 6 months or palliative)	1

(Adapted from Wells et al. 2000)
Total Score of more than 4 = Pulmonary embolism likely
Total Score of less than or equals 4 = Pulmonary embolism less likely

activity or long term immobilization—all of this need to be considered carefully as part of the risk assessment. Trained physician need to be able to assess and guide patients about the risk; preventive non-pharmacological or pharmacological measures can then be taken as appropriate for the individual depending upon the risk assessment outcome (Trunk et al. 2019).

The symptoms of DVT or PE can be assessed by Well's scoring system (Tables 6.1 and 6.2). Presently, after onset of any thrombosis like symptoms, immediate descent to low altitude region, if possible, is done to limit the symptoms and perform diagnostic tests. In case of a positive D-dimer test, a confirmatory scan (CT scan or MRI for PE or CVT; Doppler ultrasonography for DVT) is generally advised for diagnosis of thrombotic event. This practice is similar to that used for any VTE episode at plains. The travelers who have prior risk factors for VTE can be advised prophylactic oral anticoagulants—which include Factor X inhibitors (apixaban, edoxaban, or rivaroxaban) as discussed previously . Oral anticoagulant therapy is a currently accepted line of treatment, and can be started following clinical suspicion even before radiological confirmation, if there is no known bleeding risk to the patient. After excluding the surgery requirement, the oral anticoagulant therapy is continued for 6 months at the minimum with strict INR monitoring.

The prior case reports suggest that thrombogenic effect of HA exposure can go beyond the stay at HA. A report recently detailed the case of a healthy 42-year-old mountain guide who experienced shortness of breath and left calf swelling while working and skiing at 3500 m HA. Following the descent to low altitude region, he again felt shortness of breath and developed severe thoracic pain, hemoptysis, and arm cramps. The thoracic CT scan confirmed the PE in left and right lung lobes with infarction pneumonia and was put on anticoagulant therapy (Hull et al. 2016).

Due to the bleeding risk associated with anticoagulant use, there has been a need to develop a safer anti-thrombotic therapy. To address this, a highly ambitious trial of vitamin supplement was conducted for a long term of 2 years on 6000 Indian soldiers posted at HA (3500 m) (Kotwal et al. 2015). This randomized field trial was aimed at lowering the plasma homocysteine (Hcy) levels by Vitamin B12, B6, and folate and thereby reducing the thrombotic risk at HA. The supplementation was found to be protective against thrombotic episode with relative risk value of 0.29. There were total five events in the intervention group as compared to 17 events in control arm. The pro-coagulant factors—PAI-1 and fibrinogen were reduced in the intervention group (Kotwal et al. 2015). However, the exact mechanism of this effect, however, could not be established.

As hypoxemia is one of the common etiological factors in both HATE and COVID-19, and these both involve platelet activation and thrombo-inflammation, the antiplatelet therapy in COVID-19 can offer some possible insight for HATE management as well. A recent meta-analysis revealed that antiplatelet therapy against COVID-19 has not produced any significant effect on patient mortality (Hu and Song 2021), likely due to the use of aspirin as antiplatelet agent in most of those studies. The use of other antiplatelet drugs or a combination of such drugs may prove to be more beneficial. In recent times, a clinical trial was conducted, wherein an antiplatelet cocktail therapy, including GpIIb/IIIa inhibitor Tirofiban, followed by Aspirin (Cox inhibitor) and Clopidogrel (P2Y12 inhibitor) was administered to hypoxic COVID-19 patients (Viecca et al. 2020). The D-dimer and CRP levels were reduced in treated patients, indicating the anti-thrombotic and anti-inflammatory outcome of this antiplatelet regimen. Moreover, all the treated patients consistently experienced a progressive reduction in A-a O_2 gradient during the study period which indicated improved respiratory function. Although done in a few samples ($n = 5$), there was a consistent and significant improvement in the prothrombotic, inflammatory, and respiratory parameters in hypoxic COVID-19 patients. This clinical evidence supports the consideration of antiplatelet drugs and particularly that of GpIIb/IIIa and/or Clopidogrel inhibitor for HA associated thrombotic risk patients with careful monitoring of coagulation profile.

Platelet activation, however, should not only be seen as a negative prognostic factor or in simple words—a troublemaker. Platelets primarily are responsible for hemostasis and thus the inhibition of platelet activity, where the activated platelets are important for physiological events such as hemostasis, can lead to clinical mismanagement in the form of either bleeding propensity or heightened inflammation. This is because platelets do carry a pool of inflammatory and anti-inflammatory molecules, angiogenic factors, chemokines, and vasoconstrictors such as serotonin (Koupenova et al. 2018). In sepsis, for example, the platelets can serve to limit the pro-inflammatory response (Derive et al. 2012). Therefore, the efforts should be aimed at limiting the hyperactivation of platelets after confirming the platelet hyperactivity and HATE risk assessment.

6.10   Conclusion

High altitude environment is a lesser known but proven risk factor for life threatening thrombo-embolic disorders. Over the last few decades, a growing number of molecular and clinical studies have contributed to the progress in understanding the HATE pathogenesis. The underlying mechanism of HATE appears to be overlapping but still distinct from thrombosis at plains. Furthermore, in-depth studies are required to understand the biomarkers and mechanistic intricacies of this disorder. Moreover, there is a definite need for the formulation of comprehensive clinical guidelines at the global level for the prophylaxis and treatment of HATE.

Acknowledgement The authors would like to thank Dr. Aditya Arya for his kind invitation to write this chapter.

References

Ackermann M et al (2020) Pulmonary vascular endothelialitis, thrombosis, and angiogenesis in Covid-19. N Engl J Med 383(2):120–128
Agnelli G (2004) Prevention of venous thromboembolism in surgical patients. Circulation 110(24 Suppl 1):IV4–I12
Althaus K et al (2021) Antibody-induced procoagulant platelets in severe COVID-19 infection. Blood 137(8):1061–1071
Anand AC et al (2001) Thrombosis as a complication of extended stay at high altitude. Natl Med J India 14(4):197–201
Anderson FA Jr, Spencer FA (2003) Risk factors for venous thromboembolism. Circulation 107(23 Suppl 1):9–16
Bank I et al (2004) Prothrombin 20210A mutation: a mild risk factor for venous thromboembolism but not for arterial thrombotic disease and pregnancy-related complications in a family study. Arch Intern Med 164(17):1932–1937
Bartsch P, Straub PW, Haeberli A (2001) Hypobaric hypoxia. Lancet 357(9260):955–956
Basnyat B, Murdoch DR (2003) High-altitude illness. Lancet 361(9373):1967–1974
Beasley R et al (2003) eThrombosis: the 21st century variant of venous thromboembolism associated with immobility. Eur Respir J 21(2):374–376
Bendz B et al (2000) Association between acute hypobaric hypoxia and activation of coagulation in human beings. Lancet 356(9242):1657–1658

Bendz B et al (2001) Low molecular weight heparin prevents activation of coagulation in a hypobaric environment. Blood Coagul Fibrinolysis 12(5):371–374

Bergqvist D, Carlsson AS, Ericsson BF (1983) Vascular complications after total hip arthroplasty. Acta Orthop Scand 54(2):157–163

Chatterji JC et al (1982) Platelet count, platelet aggregation and fibrinogen levels following acute induction to high altitude (3200 and 3771 metres). Thromb Res 26(3):177–182

Crous-Bou M, Harrington LB, Kabrhel C (2016) Environmental and genetic risk factors associated with venous thromboembolism. Semin Thromb Hemost 42(8):808–820

Cushman M (2007) Epidemiology and risk factors for venous thrombosis. Semin Hematol 44(2): 62–69

Dahlback B (2016) Pro- and anticoagulant properties of factor V in pathogenesis of thrombosis and bleeding disorders. Int J Lab Hematol 38(Suppl 1):4–11

De Stefano V, Leone G (1995) Resistance to activated protein C due to mutated factor V as a novel cause of inherited thrombophilia. Haematologica 80(4):344–356

Delaney C et al (2021) Platelet activation contributes to hypoxia-induced inflammation. Am J Physiol Lung Cell Mol Physiol 320(3):L413–L421

Derive M et al (2012) Soluble TREM-like transcript-1 regulates leukocyte activation and controls microbial sepsis. J Immunol 188(11):5585–5592

Di Minno A et al (2020) COVID-19 and venous thromboembolism: a meta-analysis of literature studies. Semin Thromb Hemost 46(7):763–771

Dilly PN (2021) Mountain medicine. A clinical study of cold and high altitude. Michael Ward. 140 × 215 mm. Pp. 376 + x, with 21 illustrations. 1975. London: Crosby Lockwood Staples. £10. Br J Surg 63(4):333–333

Dragoman MV et al (2018) A systematic review and meta-analysis of venous thrombosis risk among users of combined oral contraception. Int J Gynaecol Obstet 141(3):287–294

Emmerich J et al (2001) Combined effect of factor V Leiden and prothrombin 20210A on the risk of venous thromboembolism--pooled analysis of 8 case-control studies including 2310 cases and 3204 controls. Study Group for Pooled-Analysis in Venous Thromboembolism. Thromb Haemost 86(3):809–816

Esmon CT (1993) Molecular events that control the protein C anticoagulant pathway. Thromb Haemost 70(1):29–35

Evans CE et al (2010) Hypoxia and upregulation of hypoxia-inducible factor 1{alpha} stimulate venous thrombus recanalization. Arterioscler Thromb Vasc Biol 30(12):2443–2451

Geerts WH et al (1994) A prospective study of venous thromboembolism after major trauma. N Engl J Med 331(24):1601–1606

Gibbs NM (1957) Venous thrombosis of the lower limbs with particular reference to bed-rest. Br J Surg 45(191):209–236

Gupta N, Ashraf MZ (2012) Exposure to high altitude: a risk factor for venous thromboembolism? Semin Thromb Hemost 38(2):156–163

Gupta N et al (2017) Activation of NLRP3 inflammasome complex potentiates venous thrombosis in response to hypoxia. Proc Natl Acad Sci U S A 114(18):4763–4768

Heit JA (2002) Venous thromboembolism epidemiology: implications for prevention and management. Semin Thromb Hemost 28(Suppl 2):3–13

Heit JA (2015) Epidemiology of venous thromboembolism. Nat Rev Cardiol 12(8):464–474

Heit JA, Spencer FA, White RH (2016) The epidemiology of venous thromboembolism. J Thromb Thrombolysis 41(1):3–14

Hisada Y et al (2015) Venous thrombosis and cancer: from mouse models to clinical trials. J Thromb Haemost 13(8):1372–1382

Hochachka PW et al (1996) Unifying theory of hypoxia tolerance: molecular/metabolic defense and rescue mechanisms for surviving oxygen lack. Proc Natl Acad Sci U S A 93(18):9493–9498

Hu Y, Song B (2021) Efficacy and safety of antiplatelet agents in patients with COVID-19 a systematic review and meta-analysis of observational studies. Res Square. https://doi.org/10.21203/rs.3.rs-417531/v1

Hudson J, Bowen A, Navia P et al (1999) The effect of high altitude on platelet counts, thrombopoietin and erythropoietin levels in young Bolivian airmen visiting the Andes. Int J Biometeorol 43:85–90. https://doi.org/10.1007/s004840050120

Hull CM, Rajendran D, Fernandez Barnes A (2016) Deep vein thrombosis and pulmonary embolism in a mountain guide: awareness, diagnostic challenges, and management considerations at altitude. Wilderness Environ Med 27(1):100–106

Jackson SP (2011) Arterial thrombosis—insidious, unpredictable and deadly. Nat Med 17(11): 1423–1436

Jha SK et al (2002) Stroke at high altitude: Indian experience. High Alt Med Biol 3(1):21–27

Jha PK et al (2018) Genome-wide expression analysis suggests hypoxia-triggered hyper-coagulation leading to venous thrombosis at high altitude. Thromb Haemost 118(7):1279–1295

Jick H et al (2000) Risk of venous thromboembolism among users of third generation oral contraceptives compared with users of oral contraceptives with levonorgestrel before and after 1995: cohort and case-control analysis. BMJ 321(7270):1190–1195

Joshi A, Jaiswal JP (2010) Deep vein thrombosis in protein S deficiency. JNMA J Nepal Med Assoc 49(177):56–58

Khorana AA et al (2007) Frequency, risk factors, and trends for venous thromboembolism among hospitalized cancer patients. Cancer 110(10):2339–2346

Kotwal J et al (2007) High altitude: a hypercoagulable state: results of a prospective cohort study. Thromb Res 120(3):391–397

Kotwal J et al (2015) Effectiveness of homocysteine lowering vitamins in prevention of thrombotic tendency at high altitude area: a randomized field trial. Thromb Res 136(4):758–762

Koupenova M et al (2018) Circulating platelets as mediators of immunity, inflammation, and thrombosis. Circ Res 122(2):337–351

Lam W, Moosavi L (2021) Physiology, Factor V. StatPearls, Treasure Island, FL

Le Joncour A et al (2020) Neutrophil-platelet and monocyte-platelet aggregates in COVID-19 patients. Thromb Haemost 120(12):1733–1735

Le Roux G et al (1992) Haemostasis at high altitude. Int J Sports Med 13(Suppl 1):S49–S51

Leiva O et al (2020) Cancer and thrombosis: new insights to an old problem. J Med Vasc 45(6S):6S8–6S16

Lippi G, Franchini M, Favaloro EJ (2009) Unsuspected triggers of venous thromboembolism—trivial or not so trivial? Semin Thromb Hemost 35(7):597–604

Lippi G, Franchini M, Targher G (2011) Arterial thrombus formation in cardiovascular disease. Nat Rev Cardiol 8(9):502–512

Lopez JA, Chen J (2009) Pathophysiology of venous thrombosis. Thromb Res 123(Suppl 4):S30–S34

Magro C et al (2020) Complement associated microvascular injury and thrombosis in the pathogenesis of severe COVID-19 infection: a report of five cases. Transl Res 220:1–13

Mahajan A et al (2019) The epidemiology of cancer-associated venous thromboembolism: an update. Semin Thromb Hemost 45(4):321–325

Maher JT, Levine PH, Cymerman A (1976) Human coagulation abnormalities during acute exposure to hypobaric hypoxia. J Appl Physiol 41(5 Pt. 1):702–707

Manne BK et al (2020) Platelet gene expression and function in patients with COVID-19. Blood 136(11):1317–1329

Mannucci PM et al (2002) Short-term exposure to high altitude causes coagulation activation and inhibits fibrinolysis. Thromb Haemost 87(2):342–343

Martinelli I et al (2003) Risk of venous thromboembolism after air travel: interaction with thrombophilia and oral contraceptives. Arch Intern Med 163(22):2771–2774

Martinod K, Deppermann C (2021) Immunothrombosis and thromboinflammation in host defense and disease. Platelets 32(3):314–324

Merli GJ, Weitz HH (2017) Venous thrombosis and cancer: what would Dr. trousseau teach today? Ann Intern Med 167(6):440–441

Middeldorp S et al (2020) Incidence of venous thromboembolism in hospitalized patients with COVID-19. J Thromb Haemost 18(8):1995–2002

Middleton EA et al (2020) Neutrophil extracellular traps contribute to immunothrombosis in COVID-19 acute respiratory distress syndrome. Blood 136(10):1169–1179

Oger E (2000) Incidence of venous thromboembolism: a community-based study in Western France. EPI-GETBP study group. Groupe d'Etude de la thrombose de Bretagne Occidentale. Thromb Haemost 83(5):657–660

Philbrick JT et al (2007) Air travel and venous thromboembolism: a systematic review. J Gen Intern Med 22(1):107–114

Pintao MC et al (2013) Protein S levels and the risk of venous thrombosis: results from the MEGA case-control study. Blood 122(18):3210–3219

Poort SR et al (1996) A common genetic variation in the $3'$-untranslated region of the prothrombin gene is associated with elevated plasma prothrombin levels and an increase in venous thrombosis. Blood 88(10):3698–3703

Pottier P et al (2009) Immobilization and the risk of venous thromboembolism. A meta-analysis on epidemiological studies. Thromb Res 124(4):468–476

Prabhakar A et al (2019) Venous thrombosis at altitude presents with distinct biochemical profiles: a comparative study from the Himalayas to the plains. Blood Adv 3(22):3713–3723

Rau JC et al (2007) Serpins in thrombosis, hemostasis and fibrinolysis. J Thromb Haemost 5(Suppl 1):102–115

Reitsma PH, Versteeg HH, Middeldorp S (2012) Mechanistic view of risk factors for venous thromboembolism. Arterioscler Thromb Vasc Biol 32(3):563–568

Rocke AS et al (2018) Thromboelastometry and platelet function during acclimatization to high altitude. Thromb Haemost 118(1):63–71

Rolfes V et al (2020) Platelets fuel the inflammasome activation of innate immune cells. Cell Rep 31(6):107615

Rosendaal FR (1999) Venous thrombosis: a multicausal disease. Lancet 353(9159):1167–1173

Schobersberger W et al (2007) Changes in blood coagulation of arm and leg veins during a simulated long-haul flight. Thromb Res 119(3):293–300

Schreijer AJ et al (2006) Activation of coagulation system during air travel: a crossover study. Lancet 367(9513):832–838

Sharma SC (1980) Platelet count on acute induction to high altitude. Thromb Haemost 43(1):24

Sharma SC (1986) Platelet count on slow induction to high altitude. Int J Biometeorol 30(1):27–32

Sharma SC et al (1977) Platelet adhesiveness in young patients with ischaemic stroke. J Clin Pathol 30(7):649–652

Simpson K (1940) Shelter deaths from pulmonary embolism. Lancet 236(6120):744

Singh I, Chohan IS (1972a) Abnormalities of blood coagulation at high altitude. Int J Biometeorol 16(3):283–297

Singh I, Chohan IS (1972b) Blood coagulation changes at high altitude predisposing to pulmonary hypertension. Br Heart J 34(6):611–617

Sueta D et al (2021) eThrombosis: a new risk factor for venous thromboembolism in the pandemic era. Res Pract Thromb Haemost 5(1):243–244

Timp JF et al (2013) Epidemiology of cancer-associated venous thrombosis. Blood 122(10):1712–1723

Toff WD et al (2006) Effect of hypobaric hypoxia, simulating conditions during long-haul air travel, on coagulation, fibrinolysis, platelet function, and endothelial activation. JAMA 295(19):2251–2261

Torgovicky R et al (2005) Sinus vein thrombosis following exposure to simulated high altitude. Aviat Space Environ Med 76(2):144–146

Trunk AD, Rondina MT, Kaplan DA (2019) Venous thromboembolism at high altitude: our approach to patients at risk. High Alt Med Biol 20(4):331–336

Tyagi T et al (2014) Altered expression of platelet proteins and calpain activity mediate hypoxia-induced prothrombotic phenotype. Blood 123(8):1250–1260

Vandenbroucke JP et al (1994) Increased risk of venous thrombosis in oral-contraceptive users who are carriers of factor V Leiden mutation. Lancet 344(8935):1453–1457

Varga EA, Moll S (2004) Cardiology patient pages. Prothrombin 20210 mutation (factor II mutation). Circulation 110(3):e15–e18

Viecca M et al (2020) Enhanced platelet inhibition treatment improves hypoxemia in patients with severe Covid-19 and hypercoagulability. A case control, proof of concept study. Pharmacol Res 158:104950

Vij AG (2009) Effect of prolonged stay at high altitude on platelet aggregation and fibrinogen levels. Platelets 20(6):421–427

Warlow C, Ogston D, Douglas AS (1976) Deep venous thrombosis of the legs after strokes. Part I—incidence and predisposing factors. Br Med J 1(6019):1178–1181

Wells PS et al (2000) Derivation of a simple clinical model to categorize patients probability of pulmonary embolism: increasing the models utility with the SimpliRED D-dimer. Thromb Haemost 83(3):416–420

Wells PS et al (2003) Evaluation of D-dimer in the diagnosis of suspected deep-vein thrombosis. N Engl J Med 349(13):1227–1235

Whayne TF Jr (2014) Altitude and cold weather: are they vascular risks? Curr Opin Cardiol 29(4): 396–402

Current Problems in Diagnosis and Treatment of High-Altitude Sickness

Gurpreet Kaur

Abstract

Acute high-altitude sickness includes various symptoms and pathology that a non-acclimatized individual experiences while travelling to higher altitudes. People travelling to high altitudes face different forms of illness (HAI) due to hypobaric hypoxia which leads to a combination of one or more syndromes including acute mountain sickness (AMS), high-altitude cerebral oedema (HACE) and high-altitude pulmonary oedema (HAPE). Lack of negligible side-effect, not so specific prophylactics and difficulties in diagnosis of acclimatization markers have led to several difficulties for the travellers. Various factors affect the occurrence of HAI that including the pace of climbing, the rate of change in altitude and genetic, environmental and epigenetic factors that code for hypoxia-inducible factors, transcription factors that affect genetic expression in response to oxygen concentrations. This chapter deals with current problems in the diagnosis of high altitude sickness and strategies that are crucial in the prevention of high-altitude related disorders. The most effective intervention for treatment to date remains the descent; however, it is not always possible to immediate descent and we need to consider other forms of interventions. In this chapter, we will also provide important insights into preventive and therapeutic approaches required for the management of various forms of high-altitude illnesses.

G. Kaur (✉)
Department of Biomedical Engineering, Indian Institute of Technology Ropar, Rupnagar, Punjab, India

Department of Biotechnology, Saraswati Group of Colleges, Mohali, Punjab, India
e-mail: gurpreet.dipas@gmail.com

© The Author(s), under exclusive license to Springer Nature Singapore Pte Ltd. 2022
N. K. Sharma, A. Arya (eds.), *High Altitude Sickness – Solutions from Genomics, Proteomics and Antioxidant Interventions*,
https://doi.org/10.1007/978-981-19-1008-1_7

Keywords

High altitude · High altitude sickness · Altitude hypoxia · Diagnosis · Treatment of HAI

7.1   Introduction

There is a keen interest in the community to travel to high altitudes for several reasons including work, entertainment, and sports or leisure activities. Upon reaching high altitudes that are more than 2500 m, some people experience various problems such as decreased partial pressure of oxygen (hypobaric hypoxia) accompanied by reduction in barometric pressure and shortness of breath. Generally, the human body can adapt itself well to hypobaric hypoxia but time is a very important factor for the acclimatization needed at high altitudes. If a person quickly ascends to the target place without taking adequate amount of time for acclimatization, high-altitude illness/sickness (HAI/HMS) happens. The best way to reduce chances of HAI is proper time for acclimatization. That means you let your body slowly get used to the changes in air pressure as you travel to higher elevations. Acclimization can last from hours to weeks, depending on both the magnitude and rate of onset of hypoxia. Acute mountain sickness (AMS) is a common, mild form of illness and usually self-limiting if managed properly. The hypobaric hypoxia activates a chain of physiological events in the body, which are tolerable in most individuals. Early symptoms of AMs manifest themselves as headache, nausea, loss of appetite, and insomnia. However, few severe cases exhibit symptoms such as shortness of breath, confusion, difficulty in walking, persistent cough, and sometimes death. However, in some cases it can lead to any of the three forms; acute mountain sickness (AMS), high-altitude cerebral oedema (HACE), and high-altitude pulmonary oedema (HAPE) (Luks et al. 2017). Various factors that affect the occurrence of HAI include pace of climbing, the rate of change in altitude, and genetic and environmental variables including genetic and epigenetic variation genes that code for hypoxia-inducible factors, transcription factors that affect genetic expression in response to oxygen concentrations (Palmer 2010). People at altitudes more than 4000 m, females, younger people, and persons with a history of severe headaches such as migraine are in high risk category group (Bartsch and Bailey 2014; Smedley and Grocott 2013).

7.2   Acute Mountain Sickness (AMS)

AMS occurs at an elevation of more than 2500 m, mostly in individuals which are not previously acclimatized with a time delay of 4–12 h post-arrival. The initial signs usually start appearing as early as reaching and spending the first night at a new altitude but can be managed by appropriate measures (Bartsch and Bailey 2014). The major symptoms of AMS include: the most prevalent headache, light-headedness, difficulty in sleeping, nausea, vomiting, insomnia, and fatigue (Palmer 2010; Luks et al. 2017). Contrary to this, HACE is an extreme form of AMs causing lethal neurologic disorders (Imray et al. 2010). If left untreated, it causes truncal ataxia and clouded consciousness and even death of the persons within 24 h. The pathophysiology of AMS remains ambiguous (Kallenberg et al. 2007) and theory of tight-fit hypothesis suggests that hypobaric hypoxia causes increase in brain volume which ultimately causes increase in intracranial pressure. This leads to impaired intracranial buffering capacity which leads to the development of severe AMS. Several factors predispose individuals to intracranial hypertension like venous hypertension/ sinovenous outflow obstruction that are clinically apparent only at elevation. Furthermore, a reduced respiratory drive in hypoxia may lead to severe deficiency of oxygen in blood (Wilson and Milledge 2008). The symptoms of AMS can be recorded on the basis of Environmental Symptoms Questionnaire and the Lake Louise AMS symptom score. Total scores are recorded as 3–5 are categorized as mild, 6–9 as moderate, and 10–12 are considered severe forms of AMS (Roach et al. 1993). There are varied responses to AMS in individuals. Some people remain unaffected by the primary symptoms while few others develop to later stages of AMS, i.e., HAPE and HACE if not given proper treatment. The non-specific nature of the AMS symptoms makes it practically difficult to identify. In the case of field diagnosis, it should be generally assumed that such non-specific symptoms are relative to AMS unless strong evidence implicates an alternative cause. Also, utmost precautions should be taken not to overlook conditions with similar non-specific presentations, e.g., hypothermia, dehydration, and hypoglycemia. An array of multiple physiological responses and anatomical factors to hypoxia such as the compensatory capacity for cerebrospinal fluid contributes to the development of pathophysiology of AMS (Luks et al. 2017). Males and females show similar susceptibility to AMS, while children are possibly less susceptible to developing AMS (Kriemler et al. 2014). People above 40–60 years are more prone to develop AMS symptoms as compared to adolescent (Honigman et al. 1993).

7.3   High-Altitude Cerebral Edema

AMS symptoms, if not resolved early, proceed to develop an acute form of disease, HACE, popularly known as the last stage of AMS (Basnyat and Murdoch 2003). This may indicate a similar pathophysiological response to both the illness (Palmer 2010). It generally takes 2 to 4 days to develop HACE symptoms after arrival at high altitude (Hall et al. 2011). Symptoms include cough, breathlessness, and decrease in exercise tolerance followed by brain swelling with edema, and intracranial hypertension (Luks et al. 2017). *Medical Society Practice Guidelines for the Prevention and Treatment of Acute Altitude Illness: 2014 Update* defined HACE as worsening cases of moderate to severe AMS displaying various neurological disorders including ataxia, extreme exhaustion, confusion, and encephalopathy and even coma. Maximal incidence of HACE is reported to be 31% (Gosainkund, Nepal, 4300 m) by Basnyat et al. (2000). An immediate pharmacological intervention to prevent or delay HACE symptoms is important in order to save the person's life. A study of severe case of HACE showed leakage in the blood–brain barrier for erythrocytes persisting for several years in the brain (Schommer et al. 2013).

More specifically, hypoxia is regarded as the major stimulator for complex reactions of events for the development of HAPE (Stream and Grissom 2008). More specifically, HAPE is caused via a "persistent imbalance between the forces that drive water into the airspace and the biologic mechanisms for its removal" (Scherrer et al. 2010). The major indicator of HAPE is pulmonary hypertension induced via hypoxia, and can be caused by any of three major pathways: a defect in the pulmonary nitric oxide synthesis pathway; increased synthesis of vasoconstrictor endothelin-1; and amplified activation of sympathetic nervous system, and sometimes defective alveolar transepithelial sodium transport (Scherrer et al. 2010). A detailed report of high altitude induced pulmonary hypertension is published by Pasha and Newman (Pasha and Newman 2010). If not treated timely, HACO can lead to coma within 24 h (Imray et al. 2010). The current situation demands urgent medical care and attention for high altitude related sickness (Bezruchka 1994; Curtis 1995; Zafren and Honigman 1997). Some reports suggest that cerebral edema can be a cause of AMS too, suggesting AMS and HACO exhibit via similar pathological processes, which often is a result of cerebral vasodilation and hyperfusion. However, some cases of severe AMS did not always show cerebral edema (Fischer et al. 2004). There are observations that report weight gain in AMS patients, suggesting abnormalities in fluid balance gain weight at altitude (Hackett et al. 1982). As of now, proper pathophysiology of AMS is unambiguous and not very clear. It may differ from HACO, but involves increased intracranial pressure, and is likely to get worse by exercise in susceptible individuals.

7.4   High-Altitude Pulmonary Oedema (HAPE)

High-altitude pulmonary edema is defined as non-cardiogenic edema that results in acute hypoxia and usually occurs in two forms (Smedley and Grocott 2013). The first one affects individuals that are previously not acclimatized and ascend quickly to heights greater than 2500 m. The second form affects people who are residing at high altitude and return to their home after a short break at lower altitude and commonly known as re-entry HAPE. It typically causes increase in the extravascular fluid in alveolar air spaces of the lung and symptoms present progressive troubled breathing, cough, sustained tiredness, decreased exercise tolerance, tachycardia, and tachypnea. It generally starts within 2 to 4 days after arrival at height, and occurs above 2500 m (Hall et al. 2011; Maggiorini 2010; Palmer 2010). The current diagnosis technique utilizes chest X-rays revealing edema in lungs. A complex cascade of events triggered by hypoxia leads to HAPE causing ataxia and altered mental status in the patients who usually suffer from preceding forms of HAI (Stream and Grissom 2008). A persistent imbalance between the forces that drive water into the airspace and the biologic mechanisms for its removal leads to development of HAPE in patients (Scherrer et al. 2010). In some advanced severe cases, chest gurgling and pink frothy sputum are also seen. Increased pulmonary vasoconstriction of small arteries and veins due to hypoxic conditions leads to expansion of the walls of blood vessel, releases the cellular junctions and hence leads to failure of alveolo-capillary membrane. Colder temperatures, fast speed of ascent (> 500 m/day), vigorous physical exercise/stress, impaired respiratory or vascular system/other contraindications are pre-disposing factors for the development of HAPE (Lobenhoffer et al. 1982). HAPO has been associated with the highest mortality rate among all HAI. 50% patients with HAPO have concomitant AMS and 14% have HACO. Study suggests that men are more susceptible to HAPE as compared to women (Maggiorini et al. 1998). Although factors such as slow ascent, oxygen, taking brief breaks, refraining from alcohol, and vigorous exercise are crucial and important, the adjunct therapeutic approaches/interventions are also very significant in the prevention of HAPE.

7.5   Prevention Strategies for Acute Mountain Sickness (AMS) and High-Altitude Cerebral Edema (HACE)

7.5.1   Non-Pharmacological Strategies

1. *Predicting who will get sick*: It would be very practical if we are able to predict the risk of development of high-altitude illness. For a person that has not had any high-altitude exposure, past experience may be considered, but not an ideal indicator of assessment for future travels. In a study, it was shown that nearly 60% of HAPE-susceptible individuals who were previously acclimatized showed progressive HAPE symptoms later on (Bärtsch et al. 2002). However, till date there are no precise methods to determine the behavior and physiology of individuals following their initial travel to high altitude. A model is already developed for screening individuals in higher risk category for any form of severe illness (HAI), but the model does not have an easy and widespread application because the model requires the individuals to breathe a hypoxic gas mixture while undergoing an exercise test in order to determine test efficacy which is highly discomforting (Canouï-Poitrine et al. 2014).

2. *Slow ascent*: The best approach in the prevention of high-altitude sickness is slow ascent while traveling to high altitude. The altitude at which an individual takes rest/sleeps is more crucial as compared to the altitude reached. At altitudes greater than 3000 m, the elevation should not be more than 500 m per day and essentially include breaks every night for at least 3 to 4 days. Due to the major inter-individual differences in their responses to high altitude sickness; some individuals can bear rapid ascent changes while others are more susceptible. Frequent travelers to high altitudes are more aware about their tolerance level and can adjust their rate of ascent accordingly. However, naïve travelers with no previous history of travel should strictly follow the limits specified. In an incident reported by the Indian Army, it was observed that 15% of the soldiers who were airlifted to extreme altitudes developed HAPE. However, none of them develop

any symptoms if acclimatized for several weeks before ascending extreme altitude (Singh et al. 1965).

3. *Pre-acclimatization*: Staged ascent which is the slow gradual ascent and staying at moderate altitudes, i.e., 2000–3100 m before traveling to high altitudes has been used traditionally to induce acclimatization in travelers. This can substantially help to overcome the symptoms of high-altitude illness, augments the protective physiological responses, and decreases the chances of development of sickness. Intermittent normobaric or hypobaric hypoxic exposures have been reported to provide beneficial role in few reports (Beidleman et al. 2004) while some studies report no clear role (Schommer et al. 2010; Fulco et al. 2011). However, implementation of this strategy with an optimal approach may be difficult for some people. One particular strategy that has been considered in recent times is resting at lower heights in an enclosed, oxygen-depleted space every night. While the efficacy of such systems remains to be confirmed clinically, few reports suggest the increased usage of this approach among travelers. Decreased incidence of AMS symptoms was reported after sleeping for 14 continuous nights in normobaric hypoxia, although there were some problems maintaining the consistent level of hypoxia (Dehnert et al. 2014). At an ascent greater than 3000 m, sleeping elevation should not be more than 500 m daily. More importantly, it is recommended not to continue to higher sleeping altitude when an individual has symptoms of HAI.

7.5.2 Other Strategies

Avoiding excessive consumption of alcohol and opiate pain drugs during ascent are some of the measures used till date for preventing acute HAI but their role is still ambiguous and debatable. Sometimes, unusual stopping of intake of caffeine in regular users may give rise to withdrawal symptoms similar to AMS (Hackett 2010). As altitude increases, there is a decrease in plasma volume. Hence, humidity level also decreases and risk of dehydration increases, but dehydration has not been directly linked to occurrence of AMS (Aoki and Robinson 1971; Castellani et al. 2010). Proper measures should be undertaken to ensure sufficient fluid intake in order to prevent dehydration taking utmost care to avoid excess of hydration at the same time. Use of diuretics, mannitol, and hypertonic saline has not been suggested in the routine preventive measure. Acetaminophen (650–1000 mg) or ibuprofen (600 mg) can be administered to relieve headache in patients as they are analgesics. However, they do not have protective effects on the pathophysiology of HACE. Other options may be budesonide, ginkgo biloba, coca leaves but due to the lack of sufficient and conflicting results, there is a need for more detailed and extensive research in this area (Zheng et al. 2014; Moraga et al. 2007; Gertsch et al. 2004). Furthermore, due to limited clinical data on efficacy, Salmeterol (beta-2 adrenergic receptor agonist drug) is not recommended in prevention of HAPE (Luks et al. 2019). Acetazolamide is the most efficient drug against development of AMS and HACE and should be considered as the first drug of choice for HAI.

7.5.3   Pharmacological Strategies

1. *Preventive strategies for AMS and HACE*: Pharmacological prophylaxis, although considered not essential always, can be initiated early in travelers based on risk evaluation for HAI. Drugs or medicines should be considered in moderate to high risk category which includes people with previous account of HAI and have an ascending rate greater than 500 m/day for altitudes greater than 2500–3000 m. Acetazolamide remains the main drug for AMS and HACE prophylaxis. It inhibits carbonic anhydrase and reduces the conversion of carbon dioxide to bicarbonate and proton, thus causing a bicarbonate diuresis and a metabolic acidosis and decrease of CSF production. A major study reported efficacy of acetazolamide at 250, 500, or 750 mg daily in the prevention of AMS. However, in a clinical setting, 250 mg on a daily basis is considered as the best working dosage (Low et al. 2012). A dose of 125 mg every 12 h (250 mg daily and 2.5 mg/kg for children) is recommended until the descent or till the time the individual reaches the high altitude and stays there for 2–3 days (McIntosh et al. 2019; Pollard et al. 2001).

   Dexamethasone is another alternative medication for individuals with intolerance or has contraindication to acetazolamide. Some studies even suggest it to be a better drug as compared to acetazolamide (Ellsworth et al. 1987). However, dexamethasone is not recommended for longer periods, typically not more than seven consecutive days due to side effects and its use must be discontinued slowly over a period (Subedi et al. 2010; Luks et al. 2014). The recommended dosage is 2 mg/6 h or 4 mg/12 h. Even though no combinatorial treatment of acetazolamide and dexamethasone is documented, combinatorial therapy can be a novel beneficial option in moderate-severe AMS patients.

   Some studies also suggested role of 1800 mg of ibuprofen per day for treatment of AMS (Lipman et al. 2012), but more studies are needed to establish its effectiveness. Additionally, risk factors with prolonged use like gastrointestinal bleeding remains unclear. Few studies also suggested the role of inhaled steroid budesonide AMS prophylaxis (Zheng et al. 2014; Chen et al. 2015). The studies point towards involvement of lungs in the underlying physiological responses associated with AMS but due to dearth of insufficient evidence and proper rationale there is no point in using this expensive drug by replacing already in-use medications (Swenson 2014).

2. *Prevention for High-altitude pulmonary edema*: A direct and apparent relationship exists between rate of ascent and development of HAPE symptoms. Therefore, gradual ascent is still the most efficacious means for prevention of all forms of HAI. Prophylactic drugs are used for prevention of HAPE.

   Dexamethasone is not precisely used for HAPE prevention, it is only prescribed in specific cases where nifedipine and tadalafil cannot be used (Maggiorini et al. 2006; Pennardt 2013). More controlled clinical studies are required to investigate the role of dexamethasone against HAPE. Nifedipine, a calcium channel blocker, is widely used therapeutic interventions for prevention of HAPE at a recommended dose of 20 mg every 8 h. It works by reducing pulmonary hypertension via constriction in pulmonary vascular resistance. Other drugs such as sildenafil and tadalafil which are phosphodiesterase inhibitors are used in some cases as and when required. Tadalafil is the more common of two because of its longer half-life in the body systems. Additionally, reports also suggest their harmful side effects (Maggiorini 2010). 10 mg of tadalafil every 12 h may be considered in case of non-responsiveness to nifedipine.

The role of acetazolamide in HAPE prophylaxis is unclear since evidence supporting its efficacy are limited, it is not recommended for HAPE. Salmeterol, an inhalable drug, when used at very high amounts (125 kg/twice per day), showed a 50% decrease in the incidence of HAPE in a more susceptible group of individuals (Sartori et al. 2002). However, the lack of sufficient clinical data with this drug limits its use but controlled use can be considered for patients with known history of frequent development of HAPE symptoms (Luks et al. 2019).

Ideally, chemoprophylaxis with any of these drugs is to be initiated on the day before they start and should be carried on till the descent or until 5 days till the traveler has spent at the altitude.

7.6 Treatment of HAI

Prevention is the most effective approach towards any illness. The substantial facts behind the current treatment of high altitude related illness and reports on their efficacy are not very accurate. Management of medical supplies and equipment at heights is also limited. Therefore, most of the time, rapid descent should be looked upon as of paramount importance and the only treatment option for HAI. It is only when immediate descent is not possible, other pharmacological and medical interventions should be considered first hand for treatment of HAI.

7.6.1 Treatment of Acute Mountain Sickness

The foremost alternative for treatment of any form of HAI is immediate descent (300–1000 m). However, during AMS, an individual's response to early symptomatic treatments can be monitored and other treatment options can be evaluated, delaying the initial descent.

AMS symptoms are generally mild and can be managed by proper rest and use of non-opiate analgesic drugs for headache and anti-emetics for gastrointestinal troubles. The efficacies of these drugs for treating AMS symptoms have not been studied in detail, but at least are known to treat mild high-altitude problems like headache and nausea (Sartori et al. 2002; Broome et al. 1994). Ibuprofen (600 mg) and acetaminophen (650–1000 mg) are two non-opioid drugs that are known to be efficacious in treating AMS. Rehydration itself cannot treat AMS, but prevent dehydration, the symptoms of which are akin to AMS. Although correlated insignificantly, proper hydration of an individual is necessary for protection against severe HAI including AMS. 250 mg of acetazolamide and dexamethasone can be viable options for treatment of severe cases of AMS who are generally unresponsive to conventional preventive methods (Harris et al. 2003; Grissom et al. 1992; Levine et al. 1989; Luks et al. 2017). Results from three controlled studies reported 8 mg as loading dose to be the most effective dosage, followed by 4 mg every 6 h by either of oral, intramuscular, or intravenous route (Ferrazzini et al. 1987; Hackett et al. 1988; Levine et al. 1989). There are no reports on the combined effect of acetazolamide and dexamethasone on HAI treatment, but combinatorial therapy can be a treatment option in severe AMS patients. Based on clinical experience, it is widely believed that supplementing oxygen at slow rates may be beneficial in reducing AMS. Therefore, some studies recommend supplemental oxygen as a treatment option for AMS in order to raise oxygen saturation more than 90% (Davis and Hackett 2017).

7.6.2   Treatment of High-Altitude Cerebral Edema

Like stated earlier, immediate descent is of utmost importance for management of HACE patients. Other treatment options should be considered when descent is not possible immediately due to certain reasons. Supplemental oxygen delivered via an oxygen tank or concentrator or via a portable hyperbaric chamber is a viable treatment approach for HACE patients (Zafren 1998). Few clinical trials have reported efficacy of portable hyperbaric chambers as an alternative to conventional options (Bärtsch et al. 1993). However, longer hours in the chamber cause several problems in patients like vomiting and claustrophobia (Davis and Hackett 2017). Additionally, recurrence of symptoms can be observed after the termination of treatment (Taber 1990).

There are a number of studies that report the use of dexamethasone in AMS patients; however, there are no studies that focus on its role in HACE patients. Sometimes dexamethasone is prescribed for HACE patients initially at a dose of 8 mg followed by 4 mg every 6 h till the descend or the complete resolution of symptoms and can be given by either of routes: oral, intramuscular or intravenous (Aksel et al. 2019). Diuretics, mannitol, hypertonic saline are not a part of standard routine treatment, nor are there any clinical reports for the use of acetazolamide in the treatment module for HACE. Furthermore, non-steroidal analgesics such as acetaminophen or ibuprofen are documented to have no major outcome on the physiological responses, but are helpful to ease mild symptoms like headache.

7.6.3   Treatment of High-Altitude Pulmonary Edema

As already stated, descent is the primary and most promising treatment approach for HAPE. However, other approaches such as supplementing oxygen, portable hyperbaric chambers, and pulmonary vasodilator agents like nifedipine and phosphodiesterase-5 inhibitors may also be beneficial. The optimal approach depends on the resources available for treatment of symptoms and it differs from case to case. For example, persons having easy access to health care facilities in their area of stay should be closely monitored at all times and given supplemental oxygen. Adequate rest and supplemental oxygen are crucial in management of HAPE. With the proper treatment approach, symptoms are manageable within hours, and complete clinical recovery is possible within a few days without the need for vasodilatory treatment in most of the cases (Zafren et al. 1996). Therefore, the use of supplemental oxygen along with proper rest is recommended to resolve the HAPE symptoms. Continuous positive airway pressure (CPAP) can be used. However, its use is limited and based on case studies (Gabry et al. 2003; Walmsley 2013). Because of several technical or clinical problems, CPAP is recommended only as a secondary treatment option, when supplemental oxygen and medications are not available. Furthermore, there is a lack of specific and systematic studies on the role of CPAP. A small random study suggested nifedipine as the preferred agent, while some other reports suggested the role of phosphodiesterase-5 inhibitors as well (Oelz et al. 1989). However, the studies lacked clinical setting and hence are limited.

30 mg nifedipine twice daily is recommended based on clinical experience. The evidence on use of phosphodiesterase-5 inhibitors is limited. However, doctors and some researchers recommend that 10 mg tadalafil or 50 mg sildenafil may be given to people in case descent is not possible or due to non-availability of nifedipine and portable hyperbaric chambers (Davis and Hackett 2017). Sildenafil, being a pulmonary vasodilator, significantly reduces the risk of a systemic low blood pressure or hypotension which is required during HAPE and does not cause systemic responses. It also increases chances of recovery from hypoxia-induced leakiness in endothelial cells. However, sildenafil has shown to aggravate AMS (Maggiorini et al. 2006). Routine administration of acetazolamide as well as dexamethasone is not indicated in HAPE because of side effects such as hypotension. Early studies also described combined application of pulmonary vasodilators, acetazolamide, inhaled β2-agonists, hyperbaric therapy in HAPE treatment (Fagenholz et al. 2007; Jones et al. 2013), but data to support this treatment approaches are limited. Severe iron deficiency has a negative impact on the pulmonary vasculature which can result into pulmonary vasoconstriction. However, it has been reported that iron supplementation has no significant protective effect against AMS/ HAPE. It is also vital to maintain an appropriate fluid balance at high altitude by preventing a state of fluid excess to avoid HAI. While some early reports described the use of diuretics (Singh et al. 1965), they are not part of routine treatment protocols for HAPE because they increase the risk of hypotension.

7.7   Better Prophylactics and Therapeutics

Upon ascent to high altitude, the resulting hypobaric hypoxia along with other factors can lead to development of multiple forms of HAI. Patients who suffer from any form of HAI can continue their journey once symptoms are completely resolved with careful consideration of pharmacological prophylaxis and slow rate of ascent. In order to treat people that cannot acclimatize, better and novel

Table 7.1 Prescribed drugs for prevention and treatment of different forms of high-altitude illness (HAI)

	Recommended dose for prevention	Recommended dose for treatment	Possible side effects
1. AMS/HACE			
(a) Acetazolamide	125 or 250 mg every 12 h	250 mg every 12 h	Adverse reactions, occurring most often early include paraesthesias, particularly "tingling in the extremities, hearing dysfunction, loss of appetite, change in taste function, nausea, vomiting and diarrhea, drowsiness
(b) Dexamethasone	2 mg every 6 h or 4 mg every 12 h	AMS:4 mg every 6 h HACE: 8 mg once; followed by 4 mg every 6 h	Hyperglycemia, retention of fluid in the body, hypokalemic alkalosis, sodium retention
2. HAPE			
Nifedipine	30 mg sustained release every 12 h	30 mg sustained release every 12 h	Headache, heat sensation, dizziness, fatigue, nausea
Tadalafil	10 mg every 12 h	10 mg every 12 h	
Salmeterol	125 µg every 12 h (inhaled)	Not recommended	

(Adapted from AM, Swenson ER, Bärtsch P. Acute high-altitude sickness. Eur Respir Rev. 2017)

prophylactics and therapeutics are urgently needed. As of today, the various pharmaceutical options require widespread medical supervision, monitoring of side effects and controlling the quantity of medications. Moreover, they are not approved for use in patients with cardiac, renal, and hepatic and other health problems. Therefore, better prophylactics/therapeutics is the need of the hour with least or no side-effects and should be safe in people suffering from other disorders (Table 7.1). The potential inclusion of alternative and newer drugs for the prevention and treatment of HAI should be prioritized. Molecules like IL-10 upregulators, corticotropin-releasing factor antagonists, glutathione S-transferase inducers may be beneficial for prophylaxis and treatment of AMS, specifically. ETA receptor antagonists, rho-kinase inhibitors, and guanylate cyclase stimulators may prove beneficial for prophylaxis and treatment of HAPE after proper research and detailed studies (Joyce et al. 2018). A new set of prophylactics, based more on natural sources may work at greater efficacy and lower doses, lowering chances of side effects and need not be prescription medicines. Studies have shown that the biomarkers and prophylactic/therapeutics from natural sources can be an alternative to unravel major current problems related to diagnosis and prophylaxis at high altitude (Paul et al. 2018). Particularly, a major protein for diagnosis of HAPE in plasma samples reported was sulfotransferase 1A1 (Sult 1A1). Sult1A1 shows highly significant differential expression in HAPE patients as compared to control groups. Other molecules like IL-1RA, HSP-70, adrenomedullin were lower in AMS-susceptible subjects throughout 10 h of hypobaric hypoxia (Paul et al. 2018). All these new biomarkers/proteins will shed light in detail regarding mechanism and diagnosis so as to make the assessment of acclimatization status during ascent to high altitude much more objective and immediate, as opposed to the non-specific and subjective scoring system which we use today.

Clinicians looking after such patients should be well trained to provide guidance and help about the behavioral and physiological changes associated with ascent to high altitude, assessment of individual's risk for high altitude related sickness, and advise proper recommendations/suggestions regarding prevention, identification of symptoms, and treatment of HAI and proper management of other medical ailments, if any. Besides, it is accepted that prevention via proper ascent rate is the best approach to all forms of HAI rather than pharmacological treatment options.

Many of the underlying aspects of HAI mechanism are still unclear, but a more focused and detailed research in this field may have greater understanding of mechanism and related symptoms for better diagnosis, management, and treatment of this serious problem.

References

Aksel G, Çorbacıoğlu ŞK, Özen C (2019) High-altitude illness: management approach. Turk J Emerg Med 19(4):121–126

Aoki VS, Robinson SM (1971) Body hydration and the incidence and severity of acute mountain sickness. J Appl Physiol 31:363–367

Bartsch P, Bailey DM (2014) Acute mountain sickness and high altitude cerebral oedema. In: Swenson ER, Bartsch P (eds) High altitude human adaptation to hypoxia. Springer, New York, pp 379–404

Bärtsch P, Merki B, Hofstetter D, Maggiorini M, Kayser B, Oelz O (1993) Treatment of acute mountain sickness by simulated descent: a randomised controlled trial. BMJ 306(6885): 1098–1101

Bärtsch P, Maggiorini M, Mairbäurl H, Vock P, Swenson ER (2002) Pulmonary extravascular fluid accumulation in climbers. Lancet 360(9332):571

Basnyat B, Murdoch DR (2003) High-altitude illness. Lancet 361(9373):1967–1974

Basnyat B, Subedi D, Sleggs J, Lemaster J, Bhasyal G, Aryal B, Subedi N (2000) Disoriented and ataxic pilgrims: an epidemiological study of acute mountain sickness and high-altitude cerebral edema at a sacred lake at 4300 m in the Nepal Himalayas. Wilderness Environ Med 11(2):89–93

Beidleman BA, Muza SR, Fulco CS, Cymerman A, Ditzler D, Stulz D, Staab JE, Skrinar GS, Lewis SF, Sawka MN (2004) Intermittent altitude exposures reduce acute mountain sickness at 4300 m. Clin Sci (Lond) 106(3):321–328

Bezruchka S (1994) Altitude illness, prevention and treatment. Cordee, Leicester

Broome JR, Stoneham MD, Beeley JM, Milledge JS, Hughes AS (1994) High altitude headache: treatment with ibuprofen. Aviat Space Environ Med 65(1):19–20

Canouï-Poitrine F, Veerabudun K, Larmignat P, Letournel M, Bastuji-Garin S, Richalet JP (2014) Risk prediction score for severe high altitude illness: a cohort study. PLoS One 9(7):e100642

Castellani JW, Muza SR, Cheuvront SN, Sils IV, Fulco CS, Kenefick RW, Beidleman BA, Sawka MN (2010) Effect of hypohydration and altitude exposure on aerobic exercise performance and acute mountain sickness. J Appl Physiol 109(6):1792–1800

Chen GZ, Zheng CR, Qin J, Yu J, Wang H, Zhang JH, Hu MD, Dong JQ, Guo WY, Lu W, Zeng Y, Huang L (2015) Inhaled budesonide prevents acute mountain sickness in young Chinese men. J Emerg Med 48(2):197–206

Curtis R (1995) Outdoor action guide to high altitude: acclimatization and illnesses. Outdoor Action Program, Princeton University, Princeton, NJ

Davis C, Hackett P (2017) Advances in the prevention and treatment of high altitude illness. Emerg Med Clin North Am 35(2):241–260

Dehnert C, Böhm A, Grigoriev I, Menold E, Bärtsch P (2014 Sep) Sleeping in moderate hypoxia at home for prevention of acute mountain sickness (AMS): a placebo-controlled, randomized double-blind study. Wilderness Environ Med 25(3):263–271

Ellsworth AJ, Larson EB, Strickland D (1987) A randomized trial of dexamethasone and acetazolamide for acute mountain sickness prophylaxis. Am J Med 83(6):1024–1030

Fagenholz PJ, Gutman JA, Murray AF, Harris NS (2007 Summer) Treatment of high altitude pulmonary edema at 4240 m in Nepal. High Alt Med Biol 8(2):139–146

Ferrazzini G, Maggiorini M, Kriemler S, Bärtsch P, Oelz O (1987) Successful treatment of acute mountain sickness with dexamethasone. Br Med J (Clin Res Ed) 294(6584):1380–1382

Fischer R, Vollmar C, Thiere M, Born C, Leitl M, Pfluger T, Huber RM (2004) No evidence of cerebral oedema in severe acute mountain sickness. Cephalalgia 24(1):66–71

Fulco CS, Muza SR, Beidleman BA, Demes R, Staab JE, Jones JE, Cymerman A (2011) Effect of repeated normobaric hypoxia exposures during sleep on acute mountain sickness, exercise performance, and sleep during exposure to terrestrial altitude. Am J Physiol Regul Integr Comp Physiol 300(2):R428–R436

Gabry AL, Ledoux X, Mozziconacci M, Martin C (2003) High-altitude pulmonary edema at moderate altitude (< 2,400 m; 7,870 feet): a series of 52 patients. Chest 123(1):49–53

Gertsch JH, Basnyat B, Johnson EW, Onopa J, Holck PS (2004) Randomised, double blind, placebo controlled comparison of ginkgo biloba and acetazolamide for prevention of acute mountain sickness among Himalayan trekkers: the prevention of high altitude illness trial (PHAIT). Br Med J 328(7443):797

Grissom CK, Roach RC, Sarnquist FH, Hackett PH (1992) Acetazolamide in the treatment of acute mountain sickness: clinical efficacy and effect on gas exchange. Ann Intern Med 116(6):461–465

Hackett PH (2010) Caffeine at high altitude: java at base cAMP. High Alt Med Biol 11:13–17

Hackett PH, Rennie D, Hofmeister SE, Grover RF, Grover EB, Reeves JT (1982) Fluid retention and relative hypoventilation in acute mountain sickness. Respiration 43:321–329

Hackett PH, Roach RC, Wood RA, Foutch RG, Meehan RT, Rennie D, Mills WJ Jr (1988) Dexamethasone for prevention and treatment of acute mountain sickness. Aviat Space Environ Med 59(10):950–954

Hall DP, Duncan K, Baillie JK (2011) High altitude pulmonary oedema. J R Army Med Corps 157(1):68–72

Harris NS, Wenzel RP, Thomas SH (2003) High altitude headache: efficacy of acetaminophen vs. ibuprofen in a randomized, controlled trial. J Emerg Med 24:383–387

Honigman B, Theis MK, Koziol-McLain J, Roach R, Yip R, Houston C, Moore LG (1993) Acute mountain sickness in a general tourist population at moderate altitudes. Ann Intern Med 118(8):587–592

Imray C, Wright A, Subudhi A, Roach R (2010) Acute mountain sickness: pathophysiology, prevention, and treatment. Prog Cardiovasc Dis 52:467–484

Jones BE, Stokes S, McKenzie S, Nilles E, Stoddard GJ (2013) Management of high altitude pulmonary edema in the Himalaya: a review of 56 cases presenting at Pheriche medical aid post (4240 m). Wilderness Environ Med 24(1):32–36

Joyce KE, Lucas SJE, Imray CHE, Balanos GM, Wright AD (2018) Advances in the available non-biological pharmacotherapy prevention and treatment of acute mountain sickness and high altitude cerebral and pulmonary oedema. Expert Opin Pharmacother 19(17):1891–1902

Kallenberg K, Bailey DM, Christ S, Mohr A, Roukens R, Menold E, Steiner T, Bärtsch P, Knauth M (2007) Magnetic resonance imaging evidence of cytotoxic cerebral edema in acute mountain sickness. J Cereb Blood Flow Metab 27(5):1064–1071

Kriemler S, Bürgi F, Wick C, Wick B, Keller M, Wiget U, Schindler C, Kaufmann BA, Kohler M, Bloch K, Brunner-La Rocca HP (2014) Prevalence of acute mountain sickness at 3500 m within and between families: a prospective cohort study. High Alt Med Biol 15(1):28–38

Levine BD, Yoshimura K, Kobayashi T, Fukushima M, Shibamoto T, Ueda G (1989) Dexamethasone in the treatment of acute mountain sickness. N Engl J Med 321(25):1707–1713

Lipman GS, Kanaan NC, Holck PS, Constance BB, Gertsch JH, PAINS Group (2012) Ibuprofen prevents altitude illness: a randomized controlled trial for prevention of altitude illness with nonsteroidal anti-inflammatories. Ann Emerg Med 59(6):484–490

Lobenhoffer H, Zink R, Brendel W (1982) High altitude pulmonary edema: analysis of 166 cases. In: High altitude physiology and medicine, pp 219–231

Low EV, Avery AJ, Gupta V, Schedlbauer A, Grocott MP (2012) Identifying the lowest effective dose of acetazolamide for the prophylaxis of acute mountain sickness: systematic review and meta-analysis. BMJ 345:e6779

Luks AM, McIntosh SE, Grissom CK, Auerbach PS, Rodway GW, Schoene RB, Zafren K, Hackett PH, Wilderness Medical Society (2014) Wilderness medical society practice guidelines for the prevention and treatment of acute altitude illness: 2014 update. Wilderness Environ Med 25(4 Suppl):S4–S14

Luks AM, Swenson ER, Bärtsch P (2017) Acute high-altitude sickness. Eur Respir Rev 26(143): 160096

Luks AM, Auerbach PS, Freer L, Grissom CK, Keyes LE, McIntosh SE, Rodway GW, Schoene RB, Zafren K, Hackett PH (2019) Wilderness medical society clinical practice guidelines for the prevention and treatment of acute altitude illness: 2019 update. Wilderness Environ Med 30(4S):S3–S18

Maggiorini M (2010) Prevention and treatment of high-altitude pulmonary edema. Prog Cardiovasc Dis 52(6):500–506

Maggiorini M, Muller A, Hofstetter D, Bartsch P, Oelz O (1998) Assessment of acute mountain sickness by different score protocols in the Swiss Alps. Aviat Space Environ Med 69(12): 1186–1192

Maggiorini M, Brunner-La Rocca HP, Peth S, Fischler M, Böhm T, Bernheim A, Kiencke S, Bloch KE, Dehnert C, Naeije R, Lehmann T, Bärtsch P, Mairbäurl H (2006) Both tadalafil and dexamethasone may reduce the incidence of high-altitude pulmonary edema: a randomized trial. Ann Intern Med 145(7):497–506

McIntosh SE, Hemphill M, McDevitt MC, Gurung TY, Ghale M, Knott JR, Thapa GB, Basnyat B, Dow J, Weber DC, Grissom K, C. (2019) Reduced acetazolamide dosing in countering altitude illness: a comparison of 62.5 vs 125 mg (the RADICAL trial). Wilderness Environ Med 30(1): 12–21

Moraga FA, Flores A, Serra J, Esnaola C, Barriento C (2007) Ginkgo biloba decreases acute mountain sickness in people ascending to high altitude at Ollagüe (3696 m) in northern Chile. Wilderness Environ Med 18(4):251–257

Oelz O, Maggiorini M, Ritter M, Waber U, Jenni R, Vock P, Bärtsch P (1989) Nifedipine for high altitude pulmonary oedema. Lancet 2(8674):1241–1244

Palmer BF (2010) Physiology and pathophysiology with ascent to altitude. Am J Med Sci 340(1): 69–77

Pasha MA, Newman JH (2010) High-altitude disorders: pulmonary hypertension: pulmonary vascular disease: the global perspective. Chest 137(Suppl 6):13–19

Paul S, Gangwar A, Bhargava K, Khurana P, Ahmad Y (2018) Diagnosis and prophylaxis for high-altitude acclimatization: adherence to molecular rationale to evade high-altitude illnesses. Life Sci 203:171–176

Pennardt A (2013) High-altitude pulmonary edema: diagnosis, prevention, and treatment. Curr Sports Med Rep 12(2):115–119

Pollard AJ, Niermeyer S, Barry P, Bärtsch P, Berghold F, Bishop RA, Clarke C, Dhillon S, Dietz TE, Durmowicz A, Durrer B, Eldridge M, Hackett P, Jean D, Kriemler S, Litch JA, Murdoch D, Nickol A, Richalet JP, Roach R, Shlim DR, Wiget U, Yaron M, Zubieta-Castillo G Sr, Zubieta-Calleja GR Jr (2001) Children at high altitude: an international consensus statement by an ad hoc committee of the International Society for Mountain Medicine, march 12, 2001. High Alt Med Biol 2(3):389–403

Roach R, Bartsch P, Hackett P, Oelz O (1993) The Lake Louise acute mountain sickness scoring system. Hypoxia Mol Med 272:4

Sartori C, Allemann Y, Duplain H, Lepori M, Egli M, Lipp E, Hutter D, Turini P, Hugli O, Cook S, Nicod P, Scherrer U (2002) Salmeterol for the prevention of high-altitude pulmonary edema. N Engl J Med 346(21):1631–1636

Scherrer U, Rexhaj E, Jayet PY, Allemann Y, Sartori C (2010) New insights in the pathogenesis of high-altitude pulmonary edema. Prog Cardiovasc Dis 52(6):485–492

Schommer K, Wiesegart N, Menold E, Haas U, Lahr K, Buhl H, Bärtsch P, Dehnert C (2010) Training in normobaric hypoxia and its effects on acute mountain sickness after rapid ascent to 4559 m. High Alt Med Biol 11(1):19–25

Schommer K, Kallenberg K, Lutz K, Bärtsch P, Knauth M (2013) Hemosiderin deposition in the brain as footprint of high-altitude cerebral edema. Neurology 81(20):1776–1779

Singh I, Kapila CC, Khanna PK, Nanda RB, Rao BD (1965) High altitude pulmonary oedema. Lancet 1(7379):229–234

Smedley T, Grocott MP (2013) Acute high-altitude illness: a clinically orientated review. Br J Pain 7(2):85–94

Stream JO, Grissom CK (2008) Update on high-altitude pulmonary edema: pathogenesis, prevention, and treatment. Wilderness Environ Med 19(4):293–303

Subedi BH, Pokharel J, Goodman TL, Amatya S, Freer L, Banskota N, Johnson E, Basnyat B (2010) Complications of steroid use on Mt. Everest. Wilderness Environ Med 21(4):345–348

Swenson ER (2014) The lungs in acute mountain sickness: victim, perpetrator, or both? Am J Med 127:899–900

Taber RL (1990) Protocols for the use of portable hyperbaric chambers for the treatment of high-altitude disorders. J Wilderness Med 1:181–192

Walmsley M (2013) Continuous positive airway pressure as adjunct treatment of acute altitude illness. High Alt Med Biol 14(4):405–407

Wilson MH, Milledge J (2008) Direct measurement of intracranial pressure at high altitude and correlation of ventricular size with acute mountain sickness: Brian Cummins' results from the 1985 Kishtwar expedition. Neurosurgery 63(5):970–974

Zafren K (1998) Gamow bag for high-altitude cerebral oedema. Lancet 352:325–326

Zafren K, Honigman B (1997) High altitude medicine. Emerg Med Clin N Am 15(1):191–222

Zafren K, Reeves JT, Schoene R (1996) Treatment of high-altitude pulmonary edema by bed rest and supplemental oxy- gen. Wilderness Environ Med 7(2):127–132

Zheng CR, Chen GZ, Yu J, Qin J, Song P, Bian SZ, Xu BD, Tang XG, Huang YT, Liang X, Yang J, Huang L (2014) Inhaled budesonide and oral dexamethasone prevent acute mountain sickness. Am J Med 127(10):1001–1009.e2

Proteomics as a Potential Tool for Biomarker Discovery

8

Vikram Dalal, Poonam Dhankhar, and Sagarika Biswas

Abstract

Proteins are a key component of the cellular orchestra, which perform diverse functions such as coordination of the molecular traffic across the cell, act as biocatalyst for biological reactions, and form the structural framework of the body. Proteomics is the main tool for proteome (the entire set of expressed protein in a given cell) study and it deals with the structural and functional characterization of a complete set of proteins present inside the cell. Knowledge derived from the proteome study can be applied for the discovery of a new biomarker candidate. In clinical research, biomarkers are the measurable indicators for the normal functioning of the cell and can indicate the presence and severity of the disease. For biomarker discovery two-dimensional electrophoresis (2DGE) and mass spectrometry are the frequently used classical methods, while other techniques such as surface-enhanced laser desorption/ionization (SELDI), two-dimensional difference gel electrophoresis (2D-DIGE), isotope-coded affinity tag (ICAT), and isobaric tag for relative and absolute quantitation (iTRAQ), etc. have been developed later. Presently advancements in technology and a better understanding of genetic knowledge speed up the process of biomarker discovery particularly in the case of tumors and autoimmune diseases. Biomarkers can give valuable information about diagnosis, prognosis, and response of a patient to treatment.

V. Dalal · P. Dhankhar
Department of Biotechnology, Indian Institute of Technology Roorkee, Roorkee, Uttarakhand, India

S. Biswas (✉)
Department of Genomics and Molecular Medicine, CSIR-Institute of Genomics and Integrative Biology, New Delhi, Delhi, India
e-mail: sagarika.biswas@igib.res.in

Keywords

Proteomics · Proteome · Autoimmune diseases · Rheumatoid arthritis ·
Biomarker · Systemic lupus erythematosus

Abbreviations

2DGE	Two-dimensional electrophoresis
2-DIGE	Two-dimensional difference gel electrophoresis
4-HHE	4-Hydroxy-2-hexenal
8-OHDG	8-Hydroxy-2′-deoxyguanosine
A1BG	A1b glycoprotein
ACTB	Beta-actin
ACTN4	Actinin Alpha 4
AGP	Alpha-1 acid glycoprotein
AGRN	Agrin
AHSG	Vimentin, alpha 2hs glycoprotein
ALB	Albumin
AMY1A,	Alpha-amylase 1
ANXA1	Annexin A1
ARG1	Arginase 1
BP1A	BPI Fold Containing Family B Member 1
CHI3L1	Chitinase-3-like protein 1
CST2	Cystatin
DCD	Dermcidin
DNA	Deoxyribose nucleic acid
ELISA	Enzyme-linked immunosorbent assay
ESI MS/MS	Nano electrospray ionization Tandem mass spectrometry
GFAP	Glial fibrillary acidic protein
GSH	Glutathione
HEL	Hexanoyl-lys adduct
HIV	Human immunodeficiency virus
HNE	4-Hydroxy 2-nonenal
HP	Haptoglobin
ICAT	Isotope-coded affinity tag
IEM	Immune electron microscopy
IGKCVIII	Ig kappa chain V-I region
IHC	Immune histochemistry
iTRAQ	Isobaric tag for relative and absolute quantitation
LC	Liquid chromatography
LCMS/MS	Liquid chromatography-mass spectrometry
MALDI	Matrix-assisted laser desorption ionization
MALDI-TOF MS	Matrix-assisted laser desorption ionization-time of flight mass spectrometry

MAST4	Microtubule-associated serine/threonine-protein kinase 4
MDA	Malondialdehyde
MRLC2	Myosin regulatory light chain
mRNA	messenger Ribose Nucleic Acid
MS	Mass spectrometry
MS	Multiple sclerosis
MUC5B	Mucin 5B, Oligomeric Mucus/Gel-Forming
MUC7	Mucin-7
NO	Nitric oxide
NO_2	Nitrogen dioxide
$ONOO^-$	Peroxynitrite
pI	Isoelectric point
PPMS	Primary progressive multiple sclerosis
PR TN3	Proteinase 3
PTMS	Post-translational modification
RA	Rheumatoid arthritis
RNAse 2	Ribonuclease
RNAse 7	Ribonuclease
RNS	Reactive nitrogen species
ROS	Reactive oxygen species
SELDI	Surface-enhanced laser desorption/ionization
SELDI-TOF MS	Surface-enhanced laser desorption/ionization-time of flight mass spectrometry
SERPINA5	Serpin Family A Member 5
SERPINA7	Serpin Family A Member
SLE	Systemic lupus erythematosus
SOD	Superoxide dismutase
SOP	Standard operating procedure
SPMS	Secondary progressive multiple sclerosis
TTR	Transthyretin
WHO	World Health Organization

8.1 Introduction

Proteomics is the scientific study of protein, encompassing their expression, structure, and functions; it deals with the characterization of the whole protein supplement of a cell, tissue, or organism (Westermeier and Naven 2002). Proteomics technology is a powerful tool for examining the total proteins expressed in an organism or a type of cell at a certain time. Proteins play a key role in the cellular processes; their expression and function vary in different cells and physiological conditions. The proteomics approach is very useful not only for fundamental research but also for applied and translational research, such as by identifying biomarkers to distinguish healthy vs diseased conditions (Faa et al. 2016).

In clinical diagnosis, several protein biomarkers are already in use, and many techniques are used for their detection and verification. Two-dimensional electrophoresis (2DGE) and mass spectrometry are the most practiced approaches for the development of protein biomarkers, in which disease samples are compared with control samples. Other techniques such as surface-enhanced laser desorption/ionization (SELDI), two-dimensional difference gel electrophoresis (2D-DIGE), isotope-coded affinity tag (ICAT), and isobaric tag for relative and absolute quantitation (iTRAQ), etc. have been developed to overcome the downside of these common approaches.

Generally, in medical science, a biomarker is a significant characteristic that shows the prognosis and severity of some disease conditions and their association with the disease is as old as medicines of the clinical laboratory. Biomarker comprises genes, variabilities at the genetic level, proteins, and changes in metabolic expression from various sources like biological fluids or tissues. The accessibility of new genetic knowledge and genomic technologies triggered the biomarker activities in recent times. There has been a constant demand for improved differential diagnoses since the number of targeted therapies has increased (Huss 2015).

8.2   How Proteomics Is Useful?

8.2.1   History of Proteomics

The early studies or researches on proteins marked the beginning of proteomics in 1975 with the invention of the two-dimensional gel techniques. O'Farrell, Klose, and Scheele were the inventor of the 2D gel technique, and they used this technique to map the proteins profile from the microorganisms such as *Escherichia coli* and higher organisms like mouse and guinea pig (O'Farrell 1975; Klose 1975; Scheele 1975; Graves and Haystead 2002). The term proteome refers to the combination of protein and genome, and Marc Wilkins, an Australian Ph.D. student, gave it in the early 1990s. He used the proteome term to describe all proteins expressed in a cell or tissue through a genome (Ademowo et al. 2013).

8.2.2   Overview of Proteomics

Proteins are the macromolecules that are an integral part of cellular life; these are the building block of the cell, which play a key role in all cellular functions and physiological processes of living organisms. There are 20 different amino acids in proteins and hundreds to thousands of which are connected in long chains to form one protein. They show remarkable variation in terms of functioning, stability, and three-dimensional structure. Proteins perform a vast array of functions within organisms such as helping to maintain the exchange of material across biological membranes, catalyze biochemical reactions, coordinate DNA, drive movement of the entire cell, etc. To date, we have entire genome sequences for many organisms even complex organisms such as human beings; however, most of the gene functions remain obscure. The twenty-first century is designated as the post-genomic or proteomic era due to the accessibility of this novel unknown ocean of data (Shruthi et al. 2016).

The term proteome refers to the sum of all genome-encoded proteins. Proteomics incorporates the analysis of a cell's or organism's overall protein composition (Clark and Pazdernik 2005). Remarkably useful information to understand different biological mechanisms, classifying cells, and tissues in disease states can be generated from proteomics data. Protein measurement methods can be exploited to analyze the structure, functions, and interactions of all proteins found in the cells and organisms. This research integrates techniques that can be used on serum and tissue to obtain valuable biological data in the form of biomarkers. It helps the physicians and researchers to understand the complex nature of their system of interest such as patients suffering from clinical issues like autoimmune diseases, cancer, and others (Posadas et al. 2005; Pradhan et al. 2010).

Proteomics is the subsequent step after genomics, for sequencing and discovering of protein pathways. On combining post-transitional amendments and splice variants, there are about ~300,000 human proteins, and this vast number makes proteomics far more complex than genomics (Nishimura et al. 2005). Originally, proteomics can be formed from both "genomics" and "proteins." Comparable with genomics where gene expression is analyzed, the expression, function, structure, and the pathway involved can be studied with proteomics. To put it on another way, valuable information like structures, identities, biochemical, and biological functions of all proteins present in a cell, organ, or organism can be generated with the help of proteomics tools (Fig. 8.1). This diverse information gathered here can further be used to distinguish between protein properties under different physiological conditions, including disease status (Kenyon et al. 2002). In this way, proteomics widens the scope for the development of clinical research and patient care (Pierce et al. 2007).

The discovery of medically relevant proteins is the main aim of clinical proteomics (Colantonio and Chan 2005; Master 2005). The data obtained through genomics combined with proteomics can provide reliable information for diagnostic purposes by identifying proteins that are expressed in the diseased state or during inflammation, trauma, and infection, including the biomarker related to these conditions (Walldius et al. 2004). Proteomics, for example, can be used to study the long-term impacts of immunosuppressive medicines by determining the site of action of a drug and the existence of its protein pathway (Traum and Schachter 2005). This approach can also be utilized for clinical analysis and drug production for specific protein pathways.

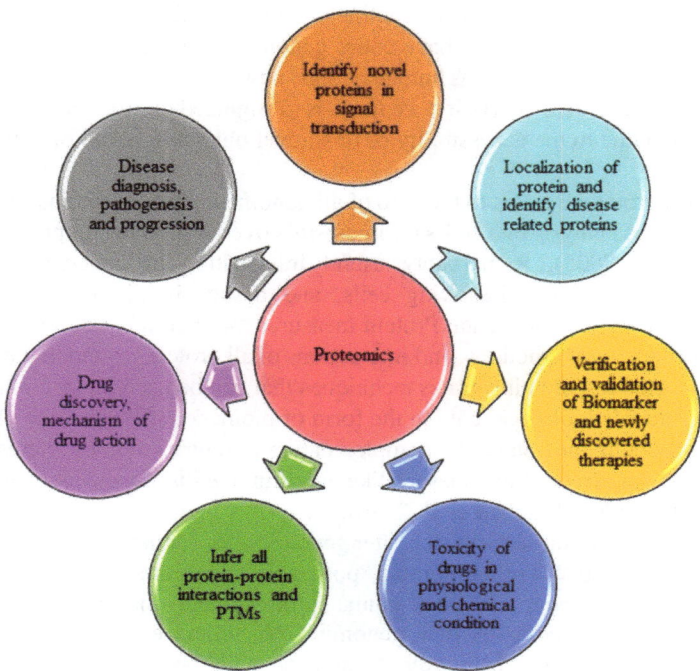

Fig. 8.1 Application of proteomics to diverse field of biological science

8.2.3 Types of Proteomics

8.2.3.1 Protein Expression Proteomics

Expression proteomics is a quantitative study of protein expressed in different specimens. This approach compares the expression proteins across the entire proteome or subproteomes between samples. With the knowledge of protein expression, the disease-related proteins can be identified, and also, the novel protein molecules can be determined, which involved in signal transduction.

8.2.3.2 Structural Proteomics

The studies mainly focus on deciphering the structural complexities of proteins that are present in complex form and in particular cellular compartment called "cell map" or structural proteomics (Blackstock and Weir 1999). Structural proteomics aims to classify and describe all the proteins present in a protein complex and cell organelle, to locate them in cell and to define all protein–protein interaction (Rout et al. 2000). The proteomic analysis has become simplified with the isolation and purification of specific subcellular organelles or protein complexes (Jung et al. 2000). Such knowledge helps to build the overall design of cells together and demonstrates how the expression of specific proteins gives a cell its distinctive features.

8.2.3.3 Functional Proteomics

Functional proteomics deals with functional characterization of protein, its activity, abundance, interaction, and also post-translational modification (Savino et al. 2012). All this information is necessary for a better understanding of protein's biological and functional role. This approach is used by many investigators for a screening of potent drug molecules to be used in different pathological conditions (Cluitmans et al. 2012; Burkhart et al. 2012). Functional characterization of proteins also provides insights about proteins signaling pathways, information about the pathogenesis of disease, and interaction of proteins with drug molecules. The data obtained from these studies produce information by characterization of a specific faction of proteins in response to signals.

8.2.3.4 Clinical Proteomics

Clinical proteomics studies mainly focused on clinical and diagnostic verification and application of newly discovered therapies and analytical markers derived from pre-clinical research performed for identification of leading molecule in correlation to drug discovery. It also involves the collection, confirmation, and standard operating procedure (SOP)-evaluation of the most appropriate and reliable process that can be incorporated into practices feasible for scientific design in clinical laboratories. Clinical proteomics approach has the potential to support almost all the known omics present in biological science including genomics, cytomics, lipidomics, metabolomics, transcriptomics, glycomics, and gene splicing variants in order to understand the advanced knowledge about the diseases and to translate this huge knowledge toward the development of new distinctive tools for a clinical practitioner (Apweiler et al. 2009).

8.2.4 Proteomics as a Platform to Discover Potent Biomarker

The discovery of a biomarker in proteomics is a stepwise process involving discovery, qualification, verification, validation, and clinical evaluation. A working group of the National Institutes of Health in 2001 defines biomarkers by referring them as quantitative indexes to assess processes of a normal biological state, pathogenic state, and pharmacological responses to a therapeutic interference (Strimbu and Tavel 2010). In addition, the United Nations and the International Labor Organization in coordination with the International Programme on Chemical Safety, led by the World Health Organization (WHO), also defined biomarkers (World Health Organization 2018). WHO presented an even broader definition that includes not only the incidence and outcome of disease but also the unintended environmental exposure of chemicals or nutrients and the effects of treatments, interventions are also taken into account in this definition (Strimbu and Tavel 2010; World Health Organization 1993).

Biomarkers can be recognized by comparing the protein profile in a typical, unaltered specimen with a diseased specimen; the characteristics of an ideal biomarker are represented in Fig. 8.2. Biomarker provides a varying degree of

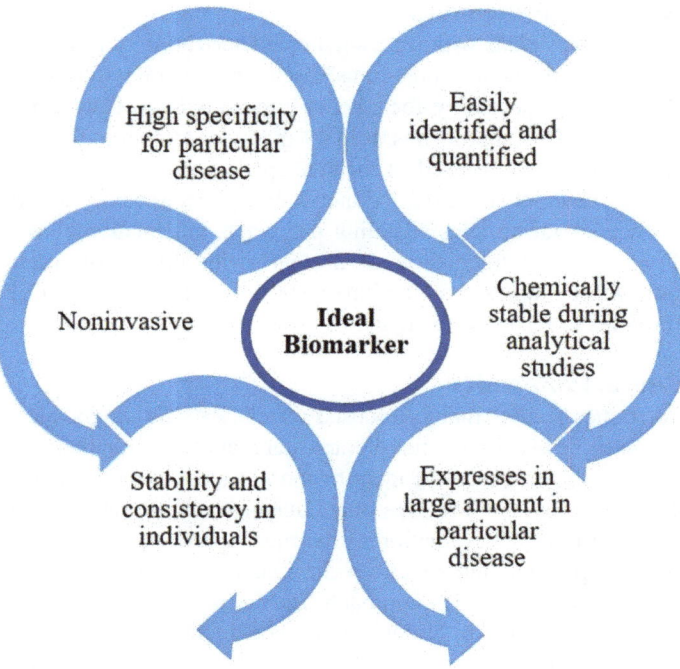

Fig. 8.2 An ideal biomarker characteristic. Proteomics is a robust technology for the discovery of new biomarkers through the assessment of global profiling of protein

information and accordingly can be classified as: (a) Diagnostic biomarker which provides information about the presence and early detection of disease, (b) Prognostic biomarkers are used to anticipate the frequency, invasion of disease, and patient reaction to certain drug treatment, (c) Predictive biomarkers generally distinguished the individuals as responding and non-responding (Gam 2012; Srinivas et al. 2002).

This distinction of the biomarker is important from the perspective of drug discovery and designing (Hamdan 2007). It has been estimated that only 2% of diseases in humans are arisen due to alteration or damage in a single gene and from non-genetic effects on gene expression but the rest 98% results from environmental effects. In this respect, proteomics can help to identify the proteins that can serve as disease-related biomarkers that contribute to disease progression. After the identification of biomarkers using mass spectrometry-based approaches, biomarkers have to be analyzed by bioinformatics and replicated in diverse populations (Amiri-Dashatan et al. 2018). Proteomics research techniques have the advantage of using biological fluid samples including plasma, urine, serum, etc., because sampling of biological fluids is simple, less invasive, and economical.

Each of these types of samples can be used for different diseases such as urine can be a valuable sample for kidney disease, since urinary proteins directly represent an alteration in kidney functions (Wu et al. 2010). The primary difficulty in biomarker discovery is heterogeneity among diseased individuals; that is why personal medication became a part of the normal treatment of cancer and other diseases. It has also been shown that the use of biomarker panels (a combination of biomarkers) gives a better diagnosis and improves sensitivity and precision for the prediction of patient response to particular drug therapy (Sarkar and Mandal 2009).

8.3   Common Proteomic Approaches

Proteomics technology is an enticing tool for the discovery of biomarkers associated with diseases as it played a significant role in the diagnosis of disease at the early stage, prognosis, and disease development (He and Chiu 2003). The discovery of proteomics biomarker is progressed in a variety of diseases such as autoimmune disease, cardiovascular disease, cancer, HIV, diabetes, renal disease, etc. A classic workflow of proteomic experiment consists of six steps in a subsequence, protein extraction, protein fractionation, comparative protein expression profiling, peptide fractionation, and mass (LCMS/MS) analysis, protein/peptide identification, protein quantification, and biochemical analysis. The common proteomic approaches are presented in Fig. 8.3.

8.3.1   Two-Dimensional Gel Electrophoresis (2DGE)

Two-dimensional gel electrophoresis (2DGE) is an efficient and reliable common proteomic technique of separating complex protein mixtures. Only 2DGE can successfully resolve ~5000 different proteins, depending on the gel size (Aslam et al. 2017). First, proteins are extracted from the tissues or cells and then subjected to the 2DGE, which separates the protein in one direction on the basis of isoelectric

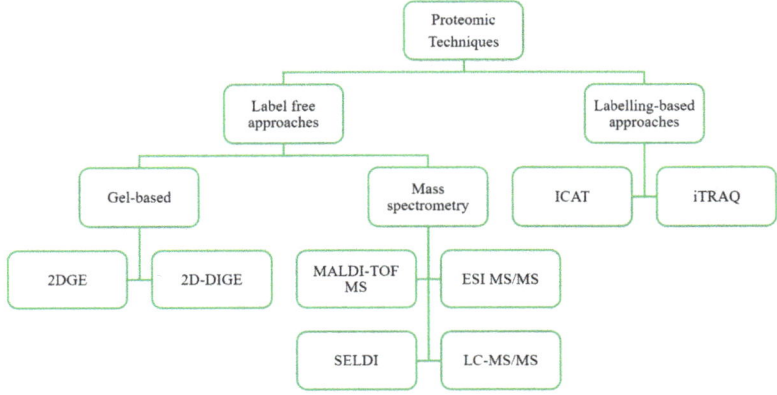

Fig. 8.3 An outline of proteomic techniques

point (pI) and in another direction on the basis of molecular weight. Following the 2DGE separation, gels are stained with Coomassie blue or silver stain to visualize the different protein spots and further, gel images were digitalized and analyzed using computer-aid 2D gel analysis software (Delyon et al. 2013). The identified interesting protein spots are applied to the tryptic in-gel digestion and the ensuing peptide mixtures are processed by the mass spectrometry (MS) to acquire the peptide fingerprints that classify the proteins by matching the genomic as well as proteomic databases. Primarily, 2DGE is used in the expression proteomics to compare quantitative and qualitative protein profiles in different samples. The emergence and disappearance of protein spots indicate the differences in the expression of protein while varying strength of spots illustrates the different levels of protein expression in certain conditions. 2DGE protein profiling is particularly important to discover the biomarkers that can compare normal to diseased samples such as tumor tissue and body fluids (He and Chiu 2003).

2DGE is much suitable to characterize the protein isoforms arising from alternative mRNA splicing, protein post-translational modifications as well as proteolytic processing because of disease or drug treatment (Castaño et al. 2006; Davidsson et al. 2002). These types of protein modification and cellular processing alter the proteins molecular weight and charge, which leads to different spot appearances in 2D gels. The 2DGE resolving ability has been widely used in structural and functional proteomic studies. But it has few downsides because it does not apply to the protein or peptides less than 10 kDa and it is also affected by the comigration issues (e.g., one stained spot can contain multiple proteins) and most notably, its applicability is limited to extremely hydrophobic proteins (Patel 2014). Nevertheless, its major failing is instability and reproducibility (Patel 2014).

8.3.2 Two-Dimensional Difference Gel Electrophoresis (2D-DIGE)

2D-DIGE was introduced by Ünlü et al. (1997) to overcome the reliability and reproducibility failure of 2DGE (O'Farrell 1975; Ünlü et al. 1997). 2D-DIGE is a modified form of 2-DGE in which control and disease samples are labeled with fluorescent Cyanide dyes (CyDyes). On a single 2-DE gel, an equal amount of labeled control and disease samples are analyzed. The respective 2-DGE patterns are easily visualized and discriminated from each other by their specific excitation and emission wavelength, which are detected using a fluorescence imager (Marouga et al. 2005). 2D-DIGE is a sensitive technique, which is able to detect up to 0.5 fmol protein. The advantages of 2D-DIGE over the 2DGE are that it reduces the experimental uncertainty and improves the data reproducibility among gels as well as precision of quantitative protein profiling, which are taken into consideration in the high-performance research environment (Minden 2007). Thus, it is a recommended technique to discover the proteomics biomarker against the disease, but the use of CyDyes made this technique quite expensive (Somiari et al. 2005; Görg et al. 2004).

8.3.3 Mass Spectrometry (MS)

Mass spectrometry (MS) is an analytical technique, which is used to obtain the structural information of protein or peptide. It produces the mass spectra of protein or peptide by measuring the mass-to-charge ratio (m/z). This method was made a core technology in proteomics by the recent improvement in MS resolution, sensitivity, and mass precision (He and Chiu 2003). The entire process of this method includes three steps. In the first step, peptides of protein are ionized using a "soft" ion source. While in the second step, ionic peptides are separated using a mass analyzer on the basis of peptides m/z values. Eventually, the separated ionic peptides are detected in a sequence by a detector and then displayed in a mass spectrum from lower to higher mass (He and Chiu 2003; Aslam et al. 2017). Commonly used MS methods consist of matrix-assisted laser desorption ionization (MALDI), nano electrospray ionization tandem MS (ESI MS/MS), and surface-enhanced laser desorption/ionization (SELDI) (Yates III 2011).

8.3.3.1 MALDI-TOF MS

The most commonly used MS method is matrix-assisted laser desorption ionization-time of flight mass spectrometry (MALDI-TOF MS) (Yates 1996, 1998). This method involves the mixing of the digested peptide with a chemical matrix and followed by the generation of ionized peptides upon the exposure of high energy laser and then resolves them by their distinct flight time into a high-vacuum tube in order to approach the detector. In the clinical microbiology laboratory, it is a new, efficient, rapid, accurate, and cost-effective method for the identification of pathogenic bacteria, as compared to the conventional techniques (Bizzini and Greub 2010; Croxatto et al. 2012; Sogawa et al. 2011). For most of the proteins, this method can accurately detect ~10 ng (200 fmol for a 50 kDa protein) (He and Chiu 2003). This is a principally used method to analyze the simple protein digests (peptide fingerprints) and it is primarily used in large-scale proteomics; unfortunately, the main concern for this method is its reproducibility (Pappin et al. 1993).

8.3.3.2 ESI MS/MS

Nano electrospray ionization tandem MS (ESI MS/MS) is another widely used MS technique in proteomics for the investigation of high molecular weight and thermally fragile polymers (Yates et al. 1997, 1999). In this method, the peptide solution is pumped into the ionization source via a microcapillary tube and resulting in the charged droplet and then vaporized in a high-vacuum chamber. The generated charged (probably multiple charged) ions are sorted based on their m/z values in the mass analyzer. This method visibly solved the issue of studying the large protein using the traditional mass spectrometry and its efficiency can be further boost by coupling it with liquid chromatography (i.e., LCMS) (El-Aneed et al. 2009). Nano electrospray ionization tandem MS technology has the advantage not only of reduced flow rates but also enhanced the ion formation mechanism (Yates et al. 1997). ESI MS has shown itself to be an effective method for rapidly characterizing biopolymers (peptides, proteins, and lipids), which can be used for the analysis of microbial species from different biomedical samples.

Throughout routine proteomics, the protein concerned is usually examined first with MALDI-TOF MS by peptide mass fingerprinting. If fingerprinting cannot recognize the proteins because of an inadequate number of proteolytic peptides or lack of an appropriate DNA database to align them confidently, then ESI MS/MS can be used for the sequencing of amino acids (He and Chiu 2003). ESI MS/MS is able to replicate the data, but it is worth noting that the relative abundance of different ions in an ESI spectra is not a true representation of the sample concentration (El-Aneed et al. 2009).

8.3.3.3 Surface-Enhanced Laser Desorption Ionization (SELDI)

In the scientific community, MALDI and ESI technologies continue to expand and form the foundation for recent ionization sources, which can be used for particular purposes (Hutchens and Yip 1993). SELDI is a newly developed proteomic approach, which combines the conventional chromatographic sample preparation with MS analysis. This approach modifies the protein chip surface chemically or biochemically to bind a particular protein group based on certain physical properties, including charge, hydrophobicity, etc. (Yip and Lomas 2002). A tiny amount of crude biological samples including serum or protein extract is spreaded on the surface of protein chip and then certain proteins that are compatible with the chemical or biochemical characteristics of the protein chip are preserved on the surface and thus isolated from the protein mixture (Yip and Lomas 2002). Further, the protein chip was mass analyzed with the ProteinChip reader that produces the mass spectra of the bound proteins (Yip and Lomas 2002). The main benefits of this technique are the ability to use a relatively small amount of crude samples and to detect the proteins of less than 6 kDa in molecular weight. Currently, it is optimized for cancer diagnosis and biomarker identification in a variety of diseases (Huang et al. 2009; Semmes et al. 2005; El-Aneed and Banoub 2006).

8.3.3.4 "Short-Gun" Approach

"Short-Gun" or LCMS/MS is a new approach, which focused on the identification and quantification of entire proteins (McCormack et al. 1997; Peng and Gygi 2001). This approach employs the reversed-phase liquid chromatography to distinct the tryptic digests of whole protein and then followed by the online ESI tandem mass spectrometry for peptide sequencing. With this method, the total protein extract from cells or tissues is pre-fractionated to simplify the protein mixture using an anion exchange column. The selected protein fractions are exposed to proteolytic digestion, and the final digestion mixture is used for liquid chromatography separation. The generated mass spectra of whole proteins are thoroughly evaluated and the protein identification is carried out through the assignment of peptides and search for databases. This method provides less manipulation of samples and mapping of total protein peptides (He and Chiu 2003). More importantly, it offers unique advantages in the discovery of protein biomarkers as compared to the other methods and it provides the high-throughput qualitative and quantitative analysis of protein

(Diamandis 2004; Wang et al. 2016). It can also be used to monitor hundreds of proteins precisely in the desired manner at the same time (Wang et al. 2016). A recent study combining this technology with a "top-down" approach significantly improves the protein recognition dynamic range and self-assurance (VerBerkmoes et al. 2002).

8.3.4 Isotope-Coded Affinity Tag (ICAT)

This is a recently introduced qualitative approach for proteome study to identify the protein-containing cysteine (Gygi et al. 1999). It involves an isotopically coded linker, cysteine sulfhydryl reactive chemical group, and a biotin tag for affinity purification. With this approach, cysteine-containing proteins are labeled with isotopic tag and then combined for proteolytic digestion. Further, the digested peptide mixture is separated by means of affinity chromatography prior to using biotin tagged to mass spectral determination (Wang et al. 2016). ICAT has been commonly used for the proteome analysis of cerebrospinal fluid to examine the mechanisms, which linked to neuropsychiatric disorders such as Alzheimer's disease (Zhang et al. 2005). Differential breast cancer profiling using the ICAT labeling technique followed by tandem mass spectrometry enabled to identify the biotinidase as a breast cancer marker (Thelen and Peck 2007). Besides this, limitations of this approach are that it is only restricted for the proteins that contain cysteine, but many proteins do not have this amino acid and the other is that it allows comparative studies simultaneously of only two conditions (Thelen and Peck 2007). However, few of the ICAT limitations may be resolved by a further advanced proteomic technique such as iTRAQ, as mentioned below (Patel 2014).

8.3.5 Isobaric Tag for Relative and Absolute Quantitation (iTRAQ)

iTRAQ is an isobaric labeling method coupled with 2D-LCMS/MS to determine the protein from different samples in a single experiment. The proteolytic peptides are labeled with iTRAQ reagent at N-terminus and amine group side chains (Ross et al. 2004). The iTRAQ reagents are of isobaric mass design, which reduces the probability of peak overlapping by differentially labeled peptides as single peaks in MS spectra. As four different iTRAQ reagents are available, a set of two or four samples can be examined in a single MS scan in addition to the identification and quantification of all peptides. The major downside of this approach is the digestion of protein into peptide before labeling, which increases the complexity of the sample. This approach has been used to discover the protein biomarker for particular diseases (Ernoult et al. 2009; Tonack et al. 2009). Over the last few years, it has been used in neuropsychiatric disease models for the comparative study of the cerebrospinal fluid proteome (Abdi et al. 2006).

8.4   Biomarker and Disease

The discovery of a biomarker in proteomics is a stepwise process involving discovery, qualification, verification, validation, and clinical evaluation (Kabeerdoss et al. 2015). Biomarkers associated with disease indicate the likely effect of treatment on the patient (risk indicator, predictive biomarkers), if disease exists previously (diagnostic biomarker), or whether the disease can progress in an individual irrespective of the type of treatment (prognostic biomarker) (Huss 2015). Proteomics techniques are also exploited to study the pathophysiology of diseases and the development of antigen-specific therapies for the autoimmune disease (Kabeerdoss et al. 2015). Presently many biomarkers have validated and verified clinically to use in medical science and are listed in Table 8.1.

Table 8.1 Clinically validated biomarker identified by quantitative proteomics method

Biomarker	Specimen	Disease	Ref.
MAST4 (microtubule-associated serine/threonine-protein kinase 4), ALB (albumin), AGRN (agrin), ANXA1 (annexin A1)	Urine	Breast cancer	Beretov et al. (2015)
ARG1 (arginase 1), AMY1A (alpha-amylase 1)	Urine	Congenital obstructive nephropathy	Lacroix et al. (2014)
RNAse2 (ribonuclease)	Urine	Ovarian cancer	Ye et al. (2006)
ACTN4 (actinin alpha 4)	Vaginal fluid	Cervical cancer	Van Raemdonck et al. (2014a)
PRTN3 (proteinase 3), SERPINA5 (serpin family A member 5)	Vaginal fluid	HIV-infection	Van Raemdonck et al. (2014b)
DCD (dermcidin)	Sweat	Atopic dermatitis	Rieg et al. (2005)
RNAse 7 (ribonuclease), DCD (dermcidin)	Sweat	Ectodermal dysplasia	Burian et al. (2015)
MUC5B (mucin 5B, oligomeric mucus/gel-forming), BPIFB1 (BPI fold containing family B member 1), CHI3L1 (chitinase-3-like protein 1)	Nasal secretion	Chronic rhinosinusitis	Saieg et al. (2015)
ACTB (beta-actin), MRLC2 (myosin regulatory light chain)	Salivary	Malignant lesions	De Jong et al. (2010)
MUC7 (mucin-7)	Salivary	Rheumatoid arthritis	Flowers et al. (2013)
SERPINA3 (serpin family A member 3)	Tear	Multiple sclerosis	Salvisberg et al. (2014)
ALB (albumin), CST2 (cystatin), IGKCVIII (Ig kappa chain V-I region)	Tear	Blepharitis	Koo et al. (2005)

8.4.1 Autoimmune Disease

Out of all human diseases, approximately 3% to 5% comes under the category of autoimmune disease. These diseases are generally heterogeneous in their medical appearance, the progression of disease, and outcomes (Hueber and Robinson 2009). There is a considerable demand for improved diagnostic techniques, the prognosis of disease and inventive therapies for better treatment of complicated autoimmune diseases such as rheumatoid arthritis (RA), diabetes mellitus (type 1 diabetes), systemic lupus erythematosus (SLE), and more. Autoimmune diseases are inflammatory conditions, which are systemic and organ-specific that involve the destruction of self-tissue by the cell-mediated immune response. It is well known that autoantibodies are present in systemic as well in tissue fluid in such disease. These types of conditions arise when the immune system reacts to self-molecule such as self-antigen autoantigen as non-self and attacked them by producing autoantibodies (Ademowo et al. 2013).

8.4.1.1 Rheumatoid Arthritis (RA)

Rheumatoid arthritis (RA) chronic illness of joints is characterized by joint destruction, loss of functionality, injury, and untimely mortality (Traum and Schachter 2005; Smolen and Aletaha 2004; Park et al. 2015). Autoantigens have been determined in the synovial fluid of the RA patients. Western blot analysis shows that the expression of a1B glycoprotein (A1BG), vimentin, alpha 2HS glycoprotein (AHSG), and glial fibrillary acidic protein (GFAP) were higher as compared to control (Mevorach 2003). A1BG and GFAP were further confirmed by enzyme-linked immunosorbent assay (ELISA), and they may be utilized as biomarkers for RA diagnosis. Transthyretin (TTR) has been identified and confirmed by western blot, ELISA, immune histochemistry (IHC), and immune electron microscopy (IEM) in the plasma of RA patients (Sharma et al. 2014). The expression of transthyretin (TTR) is more in RA as compared to normal control. It is expressed differentially in plasma, synovium, and synovial fluid samples of RA. Antibodies against TTR may be utilized for the diagnosis of RA (Hadjigogos 2003; Dalal and Biswas 2019; Dalal et al. 2017). In plasma of RA patients, haptoglobin (Hp) and alpha-1 acid glycoprotein (AGP) expressed 1.6 and 2.5 times higher as compared to a normal person (Saroha et al. 2011). Hp and AGP in combination with other diagnostic methods can be used to improve the diagnostic of RA (Saroha et al. 2011).

The level of expression of Ficolin3 is more in the plasma samples of RA as compared to the control healthy person (Roy et al. 2013). H-ficolin or ficoline3 plays a major role in innate immunity by activating the complement system. In the serum of RA patients, aglycosylated form of the level of Ig increases. The regulation of ficolin3 was determined to be related to the pathogenesis of RA. Thus, ficoline3 may be utilized as a biomarker for the RA diagnosis. Out of the nine differential expressed proteins in RA, some of the proteins like hepatoglobulin beta chain, apolipoproteins, and albumin are found to be directly involved in the development of RA.

8.4.1.2 Systemic Lupus Erythematosus (SLE)

SLE is characterized by the formation of autoimmune antibodies against nuclear components. Free radicals production can cause oxidative stress. Electron spin resonance can detect the ROS/RNS. The products of the lipids peroxidation are the biomarkers of oxidative stress in SLE (Mohan and Das 1997; Spengler et al. 2014). Catalysis of lipid peroxidation produced various products like isoprostanes, MDA, and Hexanoyl-Lys adduct (HEL) (Cracowski et al. 2002; Uchida 2003). It has been reported that 8-isoPGF2, MDA, and HNE are the well-known biomarkers for SLE (Frostegård et al. 2005; Lopez et al. 2005). The peroxidation of polyunsaturated fatty acids produces MDA. The reactions of free radical with polyunsaturated fatty acids (linolenic acids, arachidonic acid, and linoleic acid) produce the toxic aldehyde called HNE (4-hydroxy 2-nonenal) (Parola et al. 1999). Free radical peroxidation of arachidonic acid produced the F2-isoprostanes (Morrow et al. 1998). Protein oxidation biomarkers of SLE are protein nitrotyrosine and protein carbonyls (Shacter 2000). RNS species such as nitrogen dioxide (NO_2) and peroxynitrite ($ONOO^-$) react with tyrosine and produce protein nitrotyrosine. It has been found that increment in the levels of 3-nitrotyrosine is correlated with cardiac, arthritis, and renal movements in SLE patients (Ahsan 2013; Zhang et al. 2010). Oxidation of threonine, lysine, histidine, proline, and arginine can produce the protein carbonyls (Dalle-Donne et al. 2006). Several reports show the co-relation of elevation of the level of protein carbonyls with SLE (Morgan et al. 2007, 2009; Zhang et al. 2008). Oxidation of base-pair guanine of DNA forms the 8-hydroxy-2′-deoxyguanosine (8-OHdG), which is a DNA biomarker of SLE (Dalle-Donne et al. 2003; Evans et al. 2000; Zaremba and Olinski 2010). It has been reported that several other biomarkers such as GPx, SOD, CAT, Fe, GSH, Zn, ascorbic acid, selenium, and GSSG can be used for the detection of SLE (Shah et al. 2011, 2013a, b).

8.4.1.3 Type 1 Diabetes (Diabetes Mellitus)

It is a chronic autoimmune disease in which the pancreas produces little or no insulin. It is characterized by advancing dysfunctioning and the destruction of pancreatic beta cells of islets of Langerhans. In diabetic hyperglycemia, the production of free radicals can cause protein glycation and oxidative generation. Glycated hemoglobin and fructosamine levels can be used as biomarkers to detect the protein glycation in diabetes. Myeloperoxidase converts L-tyrosine to 3,3-dityrosine which can be used as a biomarker for protein oxidation in diabetes (Ylä-Herttuala 1999). It has been shown that polyunsaturated fatty acids of the cell membranes are easily attacked by free radicals (Butterfield et al. 1998). In diabetes, lipid peroxidation formed the malondialdehyde (MDA), which can react with thiobarbituric acid (Esterbauer et al. 1991). The level of vitamin E increased or decreased in diabetes. Deficiency of catalase can cause pancreas beta cells to produce ROS, which can result in the dysfunction of beta cells and diabetes (Jamieson et al. 1986).

8.4.1.4 Multiple Sclerosis

It is a multifactorial autoimmune disease in which inflammation and neurodegeneration occur together in the brain and spinal cord. As oxidative stress is a major player in the progression of MS, several compounds can be potential candidates for biomarker study. One such biomarker may be elevated levels of 8-iso-PGF2α (a marker of lipid peroxidation in vivo) in the urine of MS patients. The content of 8-iso-PGF2α in urine was found to be significantly higher in individuals with SPMS and PPMS than in control group (Jamieson et al. 1986). Destruction of lipids and cleavage of fatty acid chains lead to the formation of highly reactive aldehydes like acrolein 4-hydroxy-2-nonenal (HNE), malondialdehyde (MDA), and 4-hydroxy-2-hexenal (4-HHE) which have a longer half-life than ROS so that they can be used as markers of MS.

8.5 Conclusion

Proteomics is study of protein including expression, structure, and function of the that focuses on the entire proteins characterization present in a cell, tissue, and organism. As protein plays a leading role in cellular function. Proteomics approach has been widely used in medical science for the discovery of biomarkers, which can differentiate between normal vs diseased cells or tissue and also provide information about the prognosis of diseases. To date, many medically relevant biomarkers are in use for various diseases; however, there is a constant need for better differential diagnoses, this accelerates the discovery of new biomarker candidates. Biomarker discovery is a stepwise process in proteomics including discovery, qualification, verification, validation, and clinical evaluation phases. For the discovery of biomarkers, two-dimensional electrophoresis (2DGE) and mass spectrometry are the most common approach. Further, with the advancement of technology, other techniques such as 2D-DIGE, SELDI, ICAT, and iTRAQ are also developed to overcome the limitations of classical proteomics techniques. Noteworthy, a large number of biomarkers are discovered over time for different diseases by using these different techniques. Biomarker association with the diseases is as old as medicines of the clinical laboratory. Approximately 3–4% human diseases categorized as autoimmune disease, these are complex inflammatory conditions, which are heterogeneous in their medical representations. Biomarker, in this case, is a potent tool to identify complex autoimmune complex conditions such as rheumatoid arthritis (RA), diabetes mellitus (type 1 diabetes), and systemic lupus erythematosus (SLE).

References

Abdi F, Quinn JF, Jankovic J, McIntosh M, Leverenz JB, Peskind E et al (2006) Detection of biomarkers with a multiplex quantitative proteomic platform in cerebrospinal fluid of patients with neurodegenerative disorders. J Alzheimers Dis 9(3):293–348

Ademowo OS, Staunton L, FitzGerald O, Pennington SR (2013) Biomarkers of inflammatory arthritis and proteomics. In: Genes and autoimmunity: intracellular signaling and microbiome contribution. InTech, London, pp 237–267

Ahsan H (2013) 3-Nitrotyrosine: a biomarker of nitrogen free radical species modified proteins in systemic autoimmunogenic conditions. Hum Immunol 74(10):1392–1399

Amiri-Dashatan N, Koushki M, Abbaszadeh H-A, Rostami-Nejad M, Rezaei-Tavirani M (2018) Proteomics applications in health: biomarker and drug discovery and food industry. Iran J Pharmaceut Res 17(4):1523

Apweiler R, Aslanidis C, Deufel T, Gerstner A, Hansen J, Hochstrasser D et al (2009) Approaching clinical proteomics: current state and future fields of application in fluid proteomics. Clin Chem Lab Med 47(6):724–744

Aslam B, Basit M, Nisar MA, Khurshid M, Rasool MH (2017) Proteomics: technologies and their applications. J Chromatogr Sci 55(2):182–196

Beretov J, Wasinger VC, Millar EK, Schwartz P, Graham PH, Li Y (2015) Proteomic analysis of urine to identify breast cancer biomarker candidates using a label-free LC-MS/MS approach. PLoS One 10(11):e0141876

Bizzini A, Greub G (2010) Matrix-assisted laser desorption ionization time-of-flight mass spectrometry, a revolution in clinical microbial identification. Clin Microbiol Infect 16(11): 1614–1619

Blackstock WP, Weir MP (1999) Proteomics: quantitative and physical mapping of cellular proteins. Trends Biotechnol 17(3):121–127

Burian M, Velic A, Matic K, Günther S, Kraft B, Gonser L et al (2015) Quantitative proteomics of the human skin secretome reveal a reduction in immune defense mediators in ectodermal dysplasia patients. J Investig Dermatol 135(3):759–767

Burkhart JM, Vaudel M, Gambaryan S, Radau S, Walter U, Martens L et al (2012) The first comprehensive and quantitative analysis of human platelet protein composition allows the comparative analysis of structural and functional pathways. Blood 120(15):e73–e82

Butterfield DA, Koppal T, Howard B, Subramaniam R, Hall N, Hensley K et al (1998) Structural and functional changes in proteins induced by free radical-mediated oxidative stress and protective action of the antioxidants N-tert-butyl-α-phenylnitrone and vitamin E. Ann N Y Acad Sci 854(1):448–462

Castaño EM, Roher AE, Esh CL, Kokjohn TA, Beach T (2006) Comparative proteomics of cerebrospinal fluid in neuropathologically-confirmed Alzheimer's disease and non-demented elderly subjects. Neurol Res 28(2):155–163

Clark D, Pazdernik N (2005) Proteomics: the global analysis of proteins. Molecular biology. Elsevier Academic Press, Amsterdam, pp 717–744

Cluitmans JC, Hardeman MR, Dinkla S, Brock R, Bosman GJ (2012) Red blood cell deformability during storage: towards functional proteomics and metabolomics in the blood Bank. Blood Transfus 10(Suppl 2):s12

Colantonio DA, Chan DW (2005) The clinical application of proteomics. Clin Chim Acta 357(2): 151–158

Cracowski J-L, Durand T, Bessard G (2002) Isoprostanes as a biomarker of lipid peroxidation in humans: physiology, pharmacology and clinical implications. Trends Pharmacol Sci 23(8): 360–366

Croxatto A, Prod'hom G, Greub G (2012) Applications of MALDI-TOF mass spectrometry in clinical diagnostic microbiology. FEMS Microbiol Rev 36(2):380–407

Dalal V, Biswas S (2019) Nanoparticle-mediated oxidative stress monitoring and role of nanoparticle for treatment of inflammatory diseases. In: Nanotechnology in modern animal biotechnology. Elsevier, Amsterdam, pp 97–112

Dalal V, Sharma NK, Biswas S (2017) Oxidative stress: diagnostic methods and application in medical science. In: Oxidative stress: diagnostic methods and applications in medical science. Springer, Berlin, pp 23–45

Dalle-Donne I, Rossi R, Giustarini D, Milzani A, Colombo R (2003) Protein carbonyl groups as biomarkers of oxidative stress. Clin Chim Acta 329(1–2):23–38

Dalle-Donne I, Rossi R, Colombo R, Giustarini D, Milzani A (2006) Biomarkers of oxidative damage in human disease. Clin Chem 52(4):601–623

Davidsson P, Folkesson S, Christiansson M, Lindbjer M, Dellheden B, Blennow K et al (2002) Identification of proteins in human cerebrospinal fluid using liquid-phase isoelectric focusing as a prefractionation step followed by two-dimensional gel electrophoresis and matrix-assisted laser desorption/ionisation mass spectrometry. Rapid Commun Mass Spectrom 16(22): 2083–2088

De Jong EP, Xie H, Onsongo G, Stone MD, Chen X-B, Kooren JA et al (2010) Quantitative proteomics reveals myosin and actin as promising saliva biomarkers for distinguishing pre-malignant and malignant oral lesions. PLoS One 5(6):e11148

Delyon J, Mateus C, Lefeuvre D, Lanoy E, Zitvogel L, Chaput N et al (2013) Experience in daily practice with ipilimumab for the treatment of patients with metastatic melanoma: an early increase in lymphocyte and eosinophil counts is associated with improved survival. Ann Oncol 24(6):1697–1703

Diamandis EP (2004) Mass spectrometry as a diagnostic and a cancer biomarker discovery tool: opportunities and potential limitations. Mol Cell Proteomics 3(4):367–378

El-Aneed A, Banoub J (2006) Proteomics in the diagnosis of hepatocellular carcinoma: focus on high risk hepatitis B and C patients. Anticancer Res 26(5A):3293–3300

El-Aneed A, Cohen A, Banoub J (2009) Mass spectrometry, review of the basics: electrospray, MALDI, and commonly used mass analyzers. Appl Spectrosc Rev 44(3):210–230

Ernoult E, Bourreau A, Gamelin E, Guette C (2009) A proteomic approach for plasma biomarker discovery with iTRAQ labelling and OFFGEL fractionation. J BioMed Res 2010:927917

Esterbauer H, Schaur RJ, Zollner H (1991) Chemistry and biochemistry of 4-hydroxynonenal, malonaldehyde and related aldehydes. Free Radic Biol Med 11(1):81–128

Evans M, Cooke M, Akil M, Samanta A, Lunec J (2000) Aberrant processing of oxidative DNA damage in systemic lupus erythematosus. Biochem Biophys Res Commun 273(3):894–898

Faa G, Messana I, Fanos V, Cabras T, Manconi B, Vento G et al (2016) Proteomics applied to pediatric medicine: opportunities and challenges. Expert Rev Proteomics 13(9):883–894

Flowers SA, Ali L, Lane CS, Olin M, Karlsson NG (2013) Selected reaction monitoring to differentiate and relatively quantitate isomers of sulfated and unsulfated core 1 O-glycans from salivary MUC7 protein in rheumatoid arthritis. Mol Cell Proteomics 12(4):921–931

Frostegård J, Svenungsson E, Wu R, Gunnarsson I, Lundberg IE, Klareskog L et al (2005) Lipid peroxidation is enhanced in patients with systemic lupus erythematosus and is associated with arterial and renal disease manifestations. Arthritis Rheum 52(1):192–200

Gam L-H (2012) Breast cancer and protein biomarkers. World J Exp Med 2(5):86

Görg A, Weiss W, Dunn MJ (2004) Current two-dimensional electrophoresis technology for proteomics. Proteomics 4(12):3665–3685

Graves PR, Haystead TA (2002) Molecular biologist's guide to proteomics. Microbiol Mol Biol Rev 66(1):39–63

Gygi SP, Rist B, Gerber SA, Turecek F, Gelb MH, Aebersold R (1999) Quantitative analysis of complex protein mixtures using isotope-coded affinity tags. Nat Biotechnol 17(10):994–999

Hadjigogos K (2003) The role of free radicals in the pathogenesis of rheumatoid arthritis. Panminerva Med 45(1):7–13

Hamdan MH (2007) Cancer biomarkers: analytical techniques for discovery. Wiley, Hoboken, NJ

He QY, Chiu JF (2003) Proteomics in biomarker discovery and drug development. J Cell Biochem 89(5):868–886

Huang F, Clifton J, Yang X, Rosenquist T, Hixson D, Kovac S et al (2009) SELDI-TOF as a method for biomarker discovery in the urine of aristolochic-acid-treated mice. Electrophoresis 30(7):1168–1174

Hueber W, Robinson WH (2009) Genomics and proteomics: applications in autoimmune diseases. Pharmacogenom Pers Med 2:39

Huss R (2015) Biomarkers. Translational regenerative medicine. Elsevier, Amsterdam, pp 235–241

Hutchens TW, Yip TT (1993) New desorption strategies for the mass spectrometric analysis of macromolecules. Rapid Commun Mass Spectrom 7(7):576–580

Jamieson D, Chance B, Cadenas E, Boveris A (1986) The relation of free radical production to hyperoxia. Annu Rev Physiol 48(1):703–719

Jung E, Heller M, Sanchez JC, Hochstrasser DF (2000) Proteomics meets cell biology: the establishment of subcellular proteomes. Electrophoresis 21(16):3369–3377

Kabeerdoss J, Kurien BT, Ganapati A, Danda D (2015) Proteomics in rheumatology. Int J Rheum Dis 18(8):815–817

Kenyon GL, DeMarini DM, Fuchs E, Galas DJ, Kirsch JF, Leyh TS et al (2002) Defining the mandate of proteomics in the post-genomics era: workshop report:© 2002 National Academy of Sciences, Washington, DC, USA. Reprinted with permission from the National Academies Press for the National Academy of Sciences. All rights reserved. The original report may be viewed online at http://www.Nap.Edu/catalog/10209.html. Mol Cell Proteomics 1(10):763–780

Klose J (1975) Protein mapping by combined isoelectric focusing and electrophoresis of mouse tissues. Humangenetik 26(3):231–243

Koo B-S, Lee D-Y, Ha H-S, Kim J-C, Kim C-W (2005) Comparative analysis of the tear protein expression in blepharitis patients using two-dimensional electrophoresis. J Proteome Res 4(3): 719–724

Lacroix C, Caubet C, Gonzalez-de-Peredo A, Breuil B, Bouyssie D, Stella A et al (2014) Label-free quantitative urinary proteomics identifies the arginase pathway as a new player in congenital obstructive nephropathy. Mol Cell Proteomics 13(12):3421–3434

Lopez LR, Simpson DF, Hurley BL, Matsuura E (2005) OxLDL/β2GPI complexes and autoantibodies in patients with systemic lupus erythematosus, systemic sclerosis, and antiphospholipid syndrome: pathogenic implications for vascular involvement. Ann N Y Acad Sci 1051(1):313–322

Marouga R, David S, Hawkins E (2005) The development of the DIGE system: 2D fluorescence difference gel analysis technology. Anal Bioanal Chem 382(3):669–678

Master SR (2005) Diagnostic proteomics: back to basics? Oxford University Press, Oxford

McCormack AL, Schieltz DM, Goode B, Yang S, Barnes G, Drubin D et al (1997) Direct analysis and identification of proteins in mixtures by LC/MS/MS and database searching at the low-femtomole level. Anal Chem 69(4):767–776

Mevorach D (2003) Systemic lupus erythematosus and apoptosis. Clin Rev Allergy Immunol 25(1): 49–59

Minden J (2007) Comparative proteomics and difference gel electrophoresis. Biotechniques 43(6): 739–745

Mohan IK, Das U (1997) Oxidant stress, anti-oxidants and essential fatty acids in systemic lupus erythematosus. Prostaglandins Leukot Essent Fatty Acids 56(3):193–198

Morgan PE, Sturgess AD, Hennessy A, Davies MJ (2007) Serum protein oxidation and apolipoprotein CIII levels in people with systemic lupus erythematosus with and without nephritis. Free Radic Res 41(12):1301–1312

Morgan PE, Sturgess AD, Davies MJ (2009) Evidence for chronically elevated serum protein oxidation in systemic lupus erythematosus patients. Free Radic Res 43(2):117–127

Morrow JD, Scruggs J, Chen Y, Zackert WE, Roberts LJ (1998) Evidence that the E2-isoprostane, 15-E2t-isoprostane (8-iso-prostaglandin E2) is formed in vivo. J Lipid Res 39(8):1589–1593

Nishimura T, Ogiwara A, Fujii K, Kawakami T, Kawamura T, Anyouji H et al (2005) Disease proteomics toward bedside reality. J Gastroenterol 40(16):7–13

O'Farrell PH (1975) High resolution two-dimensional electrophoresis of proteins. J Biol Chem 250(10):4007–4021

Pappin DJ, Hojrup P, Bleasby AJ (1993) Rapid identification of proteins by peptide-mass fingerprinting. Curr Biol 3(6):327–332

Park Y-J, Chung MK, Hwang D, Kim W-U (2015) Proteomics in rheumatoid arthritis research. Immune Netw 15(4):177–185

Parola M, Bellomo G, Robino G, Barrera G, Dianzani MU (1999) 4-Hydroxynonenal as a biological signal: molecular basis and pathophysiological implications. Antioxid Redox Signal 1(3):255–284

Patel S (2014) Role of proteomics in biomarker discovery: prognosis and diagnosis of neuropsychiatric disorders. Adv Protein Chem Struct Biol 94:39–75

Peng J, Gygi SP (2001) Proteomics: the move to mixtures. J Mass Spectrom 36(10):1083–1091

Pierce JD, Fakhari M, Works KV, Pierce JT, Clancy RL (2007) Understanding proteomics. Nurs Health Sci 9(1):54–60

Posadas E, Simpkins F, Liotta L, MacDonald C, Kohn E (2005) Proteomic analysis for the early detection and rational treatment of cancer—realistic hope? Ann Oncol 16(1):16–22

Pradhan VD, Deshpande NR, Ghosh K (2010) Proteomic approach to autoimmune disorders: a review. Indian J Biotechnol 9:13–21

Rieg S, Steffen H, Seeber S, Humeny A, Kalbacher H, Dietz K et al (2005) Deficiency of dermcidin-derived antimicrobial peptides in sweat of patients with atopic dermatitis correlates with an impaired innate defense of human skin in vivo. J Immunol 174(12):8003–8010

Ross PL, Huang YN, Marchese JN, Williamson B, Parker K, Hattan S et al (2004) Multiplexed protein quantitation in Saccharomyces cerevisiae using amine-reactive isobaric tagging reagents. Mol Cell Proteomics 3(12):1154–1169

Rout MP, Aitchison JD, Suprapto A, Hjertaas K, Zhao Y, Chait BT (2000) The yeast nuclear pore complex: composition, architecture, and transport mechanism. J Cell Biol 148(4):635–652

Roy S, Biswas S, Saroha A, Sahu D, Das HR (2013) Enhanced expression and fucosylation of ficolin3 in plasma of RA patients. Clin Biochem 46(1–2):160–163

Saieg A, Brown KJ, Pena MT, Rose MC, Preciado D (2015) Proteomic analysis of pediatric sinonasal secretions shows increased MUC5B mucin in CRS. Pediatr Res 77(2):356–362

Salvisberg C, Tajouri N, Hainard A, Burkhard PR, Lalive PH, Turck N (2014) Exploring the human tear fluid: D iscovery of new biomarkers in multiple sclerosis. Proteomics Clin Appl 8(3–4): 185–194

Sarkar S, Mandal M (2009) Growth factor receptors and apoptosis regulators: signaling pathways, prognosis, chemosensitivity and treatment outcomes of breast cancer. Breast Cancer Basic Clin Res 3:47–60

Saroha A, Biswas S, Chatterjee BP, Das HR (2011) Altered glycosylation and expression of plasma alpha-1-acid glycoprotein and haptoglobin in rheumatoid arthritis. J Chromatogr B 879(20): 1839–1843

Savino R, Paduano S, Preianò M, Terracciano R (2012) The proteomics big challenge for biomarkers and new drug-targets discovery. Int J Mol Sci 13(11):13926–13948

Scheele GA (1975) Two-dimensional gel analysis of soluble proteins. Charaterization of Guinea pig exocrine pancreatic proteins. J Biol Chem 250(14):5375–5385

Semmes OJ, Feng Z, Adam B-L, Banez LL, Bigbee WL, Campos D et al (2005) Evaluation of serum protein profiling by surface-enhanced laser desorption/ionization time-of-flight mass spectrometry for the detection of prostate cancer: I. Assessment of platform reproducibility. Clin Chem 51(1):102–112

Shacter E (2000) Quantification and significance of protein oxidation in biological samples. Drug Metab Rev 32(3–4):307–326

Shah D, Wanchu A, Bhatnagar A (2011) Interaction between oxidative stress and chemokines: possible pathogenic role in systemic lupus erythematosus and rheumatoid arthritis. Immunobiology 216(9):1010–1017

Shah D, Sah S, Wanchu A, Wu MX, Bhatnagar A (2013a) Altered redox state and apoptosis in the pathogenesis of systemic lupus erythematosus. Immunobiology 218(4):620–627

Shah D, Sah S, Nath SK (2013b) Interaction between glutathione and apoptosis in systemic lupus erythematosus. Autoimmun Rev 12(7):741–751

Sharma S, Ghosh S, Singh LK, Sarkar A, Malhotra R, Garg OP et al (2014) Identification of autoantibodies against transthyretin for the screening and diagnosis of rheumatoid arthritis. PLoS One 9(4):e93905

Shruthi BS, Vinodhkumar P, Selvamani (2016) Proteomics: a new perspective for cancer. Adv Biomed Res 5:67

Smolen J, Aletaha D (2004) Patients with rheumatoid arthritis in clinical care. Ann Rheum Dis 63(3):221–225

Sogawa K, Watanabe M, Sato K, Segawa S, Ishii C, Miyabe A et al (2011) Use of the MALDI BioTyper system with MALDI–TOF mass spectrometry for rapid identification of microorganisms. Anal Bioanal Chem 400(7):1905

Somiari RI, Somiari S, Russell S, Shriver CD (2005) Proteomics of breast carcinoma. J Chromatogr B 815(1–2):215–225

Spengler M, Svetaz M, Leroux M, Bertoluzzo S, Parente F, Bosch P (2014) Lipid peroxidation affects red blood cells membrane properties in patients with systemic lupus erythematosus. Clin Hemorheol Microcirc 58(4):489–495

Srinivas PR, Verma M, Zhao Y, Srivastava S (2002) Proteomics for cancer biomarker discovery. Clin Chem 48(8):1160–1169

Strimbu K, Tavel JA (2010) What are biomarkers? Curr Opin HIV AIDS 5(6):463

Thelen JJ, Peck SC (2007) Quantitative proteomics in plants: choices in abundance. Plant Cell 19(11):3339–3346

Tonack S, Aspinall-ODea M, Jenkins RE, Elliot V, Murray S, Lane CS et al (2009) A technically detailed and pragmatic protocol for quantitative serum proteomics using iTRAQ. J Proteomics 73(2):352–356

Traum AZ, Schachter AD (2005) Transplantation proteomics. Pediatr Transplant 9(6):700–711

Uchida K (2003) 4-Hydroxy-2-nonenal: a product and mediator of oxidative stress. Prog Lipid Res 42(4):318–343

Ünlü M, Morgan ME, Minden JS (1997) Difference gel electrophoresis. A single gel method for detecting changes in protein extracts. Electrophoresis 18(11):2071–2077

Van Raemdonck GA, Tjalma WA, Coen EP, Depuydt CE, Van Ostade XW (2014a) Identification of protein biomarkers for cervical cancer using human cervicovaginal fluid. PLoS One 9(9): e106488

Van Raemdonck G, Zegels G, Coen E, Vuylsteke B, Jennes W, Van Ostade X (2014b) Increased serpin A5 levels in the cervicovaginal fluid of HIV-1 exposed seronegatives suggest that a subtle balance between serine proteases and their inhibitors may determine susceptibility to HIV-1 infection. Virology 458:11–21

VerBerkmoes NC, Bundy JL, Hauser L, Asano KG, Razumovskaya J, Larimer F et al (2002) Integrating "top-down" and "bottom-up" mass spectrometric approaches for proteomic analysis of Shewanella oneidensis. J Proteome Res 1(3):239–252

Walldius G, Jungner I, Aastveit AH, Holme I, Furberg CD, Sniderman AD (2004) The apoB/apoA-I ratio is better than the cholesterol ratios to estimate the balance between plasma proatherogenic and antiatherogenic lipoproteins and to predict coronary risk. Clin Chem Lab Med 42(12): 1355–1363

Wang H, Shi T, Qian W-J, Liu T, Kagan J, Srivastava S et al (2016) The clinical impact of recent advances in LC-MS for cancer biomarker discovery and verification. Expert Rev Proteomics 13(1):99–114

Westermeier R, Naven T (2002) Proteomics in practice. Wiley-VCH, Weinheim

World Health Organization (1993) International Programme on chemical safety (IPCS) Biomarkers and risk assessment: concepts and principles. World Health Organization, Geneva, p 57

World Health Organization (2018) International Programme on chemical safety. Biomarkers in risk assessment: validity and validation, 2001. WHO, Geneva

Wu J, Chen Y-d, Gu W (2010) Urinary proteomics as a novel tool for biomarker discovery in kidney diseases. J Zhejiang Univ Sci B 11(4):227–237

Yates JR (1996) Protein structure analysis by mass spectrometry. Methods Enzymol 271:351–377

Yates JR III (1998) Mass spectrometry and the age of the proteome. J Mass Spectrom 33(1):1–19

Yates JR III (2011) A century of mass spectrometry: from atoms to proteomes. Nat Methods 8(8): 633–637

Yates JR, McCormack AL, Schieltz D, Carmack E, Link A (1997) Direct analysis of protein mixtures by tandem mass spectrometry. J Protein Chem 16(5):495–497

Yates JR, Carmack E, Hays L, Link AJ, Eng JK (1999) Automated protein identification using microcolumn liquid chromatography-tandem mass spectrometry. In: 2-D proteome analysis protocols. Springer, Berlin, pp 553–569

Ye B, Skates S, Mok SC, Horick NK, Rosenberg HF, Vitonis A et al (2006) Proteomic-based discovery and characterization of glycosylated eosinophil-derived neurotoxin and COOH-terminal osteopontin fragments for ovarian cancer in urine. Clin Cancer Res 12(2):432–441

Yip T-T, Lomas L (2002) SELDI ProteinChip® array in oncoproteomic research. Technol Cancer Res Treat 1(4):273–279

Ylä-Herttuala S (1999) Oxidized LDL and atherogenesis. Ann N Y Acad Sci 874(1):134–137

Zaremba T, Olinski R (2010) Oxidative DNA damage–analysis and clinical significance. Postepy Biochem 56(2):124–138

Zhang J, Goodlett DR, Quinn JF, Peskind E, Kaye JA, Zhou Y et al (2005) Quantitative proteomics of cerebrospinal fluid from patients with Alzheimer disease. J Alzheimers Dis 7(2):125–133

Zhang Q, Ye D, Chen G (2008) Study on the relationship between protein oxidation and disease activity in systemic lupus erythematosus. Zhonghua liu xing bing xue za zhi= Zhonghua liuxingbingxue zazhi 29(2):181–184

Zhang Q, Ye D, Chen G, Zheng Y (2010) Oxidative protein damage and antioxidant status in systemic lupus erythematosus. Clin Exp Dermatol Exp Dermatol 35(3):287–294

Serum and Plasma Proteomics for High Altitude Related Biomarker Discovery

9

Aditya Arya and Amit Kumar

Abstract

Serum and plasma are widely used biological fluids for a large variety of biochemical and pathological tests which have been used since historical times for diagnosis and prophylactic analysis. High altitude physiology and related pathologies are geographically localized socio-economic issues of a large set of population. In the last few decades, considerable research has been done to explore the underlying proteomic changes in serum and plasma of humans as well as in the model organisms. Recent studies have led to the emergence of a number of potential prophylactic and therapeutic targets which enable better and timely diagnosis, effective therapy, and also susceptibility testing of high altitude related illness. As proteome is highly dynamic in time and space, and serum or plasma is known to reflect the proteomic changes in various organs. Although analysis of plasma or serum is an invasive approach, yet the specificity of plasma-based markers holds great promise in diagnostics. In this chapter, we will discuss general approaches to use plasma and serum proteome, followed by examples of potential biomarkers or indicators of hypobaric-hypoxia-induced perturbations.

Keywords

Serum · Plasma · Biomarker · High altitude pathophysiology · Hypobaric hypoxia

A. Arya (✉)
National Institute of Malaria Research, Indian Council of Medical Research, New Delhi, India

A. Kumar
Sathyabama Institute of Science and Technology, Chennai, Tamil Nadu, India

9.1   Introduction

Plasma proteomics has been a cornerstone of biomarker discovery for a large number of physiological and pathological conditions since historical times. The use of plasma proteins for the diagnosis and assessment of various organ function tests is popular enough across clinicians. Although diseases and pathological conditions show less wide alteration in the proteins compared to altered physiological conditions, hence it is challenging to establish a bio-molecular marker in physiological conditions compared to the pathological condition. High altitude physiology or hypobaric hypoxia is a less-known yet complex physiological condition that is characterized by alterations in the proteomic profile of various organs mediated by physiological responses. A large number of proteomic studies including basic studies using 2D gel electrophoresis or differential gel electrophoresis (DIGE) were performed in the past in the authors' lab and were later re-established using newer and more reliable approaches including targeted and quantitative proteomics. With the increasing availability of proteome spectra database and improved mass resolution in mass spectrometry, the number of proteins identified per sample has substantially increased and thus brought about the refinement of the proteomic alteration during high altitude exposure. While the key question of proteomics studies in high altitude remains elucidation of physiological malfunctions under hypobaric hypoxia, the exact chronology of molecular events and physiological changes is yet to be established. Nevertheless, adaptability to high altitudes especially by comparing natives and lowlander travelers is yet another prospective area under investigation. Defense establishments have an advantage in gathering organized study subjects due to homogenous age-matched and diet-matched individuals, which is otherwise not possible to attain. In more than a dozen studies, thousands of differentially expressed proteins have already been deciphered to follow a pattern and therefore with a potential to be established as a biomarker either for high altitude sickness susceptibility or indicators of hypobaric-hypoxia-adaptability. This chapter mainly focuses on various proteomic approaches using plasma/serum as samples. Some of the studies by authors and other potential biomarkers are highlighted in the end.

9.2   Plasma and Serum and Their Potentials in the Diagnosis

Serum and plasma both are blood fractions in liquid form which are free from any type of cells. However, the serum is obtained after clotting the blood, hence it is free from clotting factors, while plasma is obtained from unclotted blood (by collecting blood in an anticoagulant medium) and hence richer in terms of proteins. The protein concentration in plasma/serum is approximately 60–80 mg/mL and around 50–60% proportion is occupied by some of the most abundant proteins, namely albumins and 40% globulins, in particular, 10–20% are immunoglobulin G, the most prominent immunoglobulin (Leeman et al. 2018), rest 40–50% are other proteins, some of which have very low abundance. In the early days when proteomics approaches were not advanced enough, it was difficult to study the low abundance proteins and enumerate the diversity of plasma proteins, but now with the help of high-resolution mass spectrometry (HR-MS), several studies have independently enumerated the profiling of serum/plasma proteins to about 15,000, encoded by more than 12,000 genes (Nedelkov et al. 2005). Present-day proteomics methods also provide techniques for sequestering and thus eliminating the high abundance proteins by affinity-based approaches making the identification of low abundance proteins easier. Besides the canonical plasma and serum proteins, which play roles in hemodynamics and other plasma functions, several other non-canonical proteins are often secreted from various organs and therefore reveal the pathophysiological state of various organs. In fact, the use of isoenzymes, limited to specific organs, has been historically used for organ function tests. More recently, the availability of high throughput quantitative proteomics approaches, such as iTRAQ, MRM, TMT, etc. has provided opportunities to researchers to explore the minute differences in protein levels in various physiological or pathological conditions. These advancements have certainly empowered the proteomic biomarker discovery pipelines and several prospective proteins are increasingly being described as potential biomarkers for pathological or even physiological alterations. Plasma or serum is considered a good alternative for tissue or biopsy sampling for proteomic studies, especially for human studies, due to their less invasive nature; however, other samples such as urine or saliva are completely non-invasive but low diversity and low abundance of proteins in those samples is still a challenge and keeps the importance of plasma and serum relevant in the pathophysiological context.

9.3   Proteomics Approaches to Study Plasma and Serum

Serum preparation requires the collection of blood and separation of serum by placing the blood-containing tube in the ice for an hour after collection for coagulation of blood (coagulation can be hastened by adding agents like thrombin), followed by centrifugation. The supernatant after the centrifuge contains serum proteins. Plasma separation involves the addition of coagulants, e.g. EDTA, heparin followed by centrifugation. Serum and plasma contain several abundant proteins such as albumin and immunoglobulin, which sometimes need to be removed using commercially available affinity columns if the study is aimed to investigate lesser abundant proteins.

Isolation of the proteins from serum and plasma begins with cell disruption, e.g. sonication, followed by precipitation, and finally, purification to remove even the slightest impurities, for proteomics. There are several methods for the precipitation of proteins such as the use of cosmotropic agents (urea) which breaks the hydrogen bond between molecules and denature it, chaotropic agents (ammonium sulfate) which promote the formation of water–water hydrogen bonds, and protein losses hydration and precipitates out, a phenomenon known as salting out. Ionic precipitation methods of precipitation use acids like trichloroacetic acid, salicylic acid. Proteins can also be precipitated by organic solvents such as acetone, ethanol, etc. based on their potential to dehydrate and remove the water of hydration. After precipitation of protein, removal of salts and other impurities needs to be achieved through methods such as ion-exchange chromatography or dialysis, before downstream proteomic processes. One must take precautions in sample collection, handling, and storage of the samples and standardization of the proteome extraction method; otherwise, it can affect the reproducibility of the results (Blonder et al. 2008). Several proteomics approaches are available, such as gel-based including one-dimensional and 2D polyacrylamide gel electrophoresis, and gel-free including label-free and labeled such as isotope-coded affinity tag (ICAT), isobaric tragic for relative and absolute quantification (iTRAQ) shotgun proteomics, etc.

9.3.1   Immunoblotting

Immunoblotting is a rapid assay for the detection of target proteins that works by exploiting specificity inherent in antigen–antibody recognition. The step involves electrophoretic separation of proteins, followed by transfer and bindings to the nitrocellulose/nylon/PVDF membrane, treatment with selective primary and a secondary antibody, and visualization using chromogenic or chemiluminescent substrates (Gallagher and Chakavarti 2008). Immunoblotting has been one of the most conventional proteomic approaches in understanding effect in the high altitudes (Gangwar et al. 2020; Lopez et al. 1975).

9.3.2   2D Gel Electrophoresis and Mass Spectrometry

2D polyacrylamide gel electrophoresis is a widely used technique to separate and visualize the proteins based on their mass and charge (Aslam et al. 2017) followed by identification of selected protein spots using mass spectrometry (MS). In the 2D gel electrophoresis, the proteins are separated into two steps, first in the one dimension using the pI values and then in the second dimension based on their relative molecular weight. Upon comparison of gels from different experimental conditions, selected proteins are generally chosen for further identification through MS. MS measures the mass-to-charge ratio, thus determining the molecular weight of the proteins. MS involves three steps: the first step is to transform the peptides into gas-phase ions. The most common ionization methods include MALDI (matrix-assisted laser desorption ionization), SELDI (surface-enhanced laser desorption/ ionization), and ESI (electrospray ionization). The second step is the separation of ions on the basis of mass/charge (m/z) values in the presence of an electric/magnetic field and the final step is to measure the m/z values of each separated ion (Aslam et al. 2017). 2D gel electrophoresis and MS have provided insights into the human proteome changes at high altitude conditions (Ahmad et al. 2013).

9.3.3   Non-gel Based Quantitative Proteomics

Quantification of proteins in a sample can be mainly performed by 2D gel electro-phoresis or MS. MS can identify and quantify the changes in the protein (Matthiesen and Bunkenborg 2013). In general, quantification can be done either by labeling peptides or can be label-free. Among the labeling, the most common ones are isotope-coded affinity tag (ICAT), isobaric tragic for relative and absolute quantification (iTRAQ), stable isotope labeling by amino acids in cell culture (SILAC), isotope-coded protein labeling (ICPL). These labels are useful to study protein changes in complex samples (Veenstra 2007). The development of the label-free approach has helped achieve faster, cleaner, and simple quantification results. In this technique, different proteome samples are prepared and separated separately using LC-MS/MS or LC/LC-MS/MS followed by protein quantification. Quantification is based on two categories, first, measurement of ion intensity changes (such as peptide peak/peak heights) in chromatography, and second is based on spectral counts for individual samples. Direct comparison between different sample analyses tells about changes in the protein abundance (Zhu et al. 2010). MS-based proteomics in high altitude stress and diseases have been carried out extensively in the last one decades on both humans and experimental animals under different conditions such as hypoxia (Gao et al. 2017).

9.4   Key Considerations for Biomarker Discovery for High Altitude Physiology

Biomarker discovery has been a long pursued domain of clinical sciences which has got attention with the advent of omics technologies. A typical dictionary definition of a biomarker is "a naturally occurring molecule, gene, or characteristic by which a particular pathological or physiological process, disease, etc. can be identified" or "a biomarker that may predict aggressive disease recurrence in liver transplant recipients" (Oxford dictionary). The standard pipeline of biomarker discovery using proteomics includes several important steps. The first and foremost consideration is about the sample choice and experimental regimens. Simulated hypoxia on experimental rats might not be physiologically identical to the actual hypobaric hypoxia observed on high altitudes. Also, the physiological responses of humans at a specific altitude are different from experimental animals such as rats. Hence, conclusions from simulated hypoxia on animal models may not be directly extrapolated for the biomarker discovery. It is therefore important that separate experimental regimens and models must be chosen to specifically define hypoxia markers.

The next important aspect in using proteomics for biomarker discovery is the nature of the proteomic approach. Classical approaches such as 2D-gel electrophoresis are now considered obsolete and non-reproducible, hence potential biomarkers identified using 2DGE must be validated enough to be used clinically or confirmed using alternative sensitive and reproducible methods. Emerging approaches such as quantitative proteomics (using iTRAQ, TMT, or label-free) have been known to show a better coverage and hence preferred over conventional mass spectrometric methods. Moreover, targeted proteomics and shotgun proteomics are also emerging approaches for biomarker discovery.

Another highly important consideration for the biomarker discovery in high altitude biology is the choice of sample. Most often plasma or serum is considered as a good choice due to its richness and easily identifiable secretary proteome of various organs. However, one must consider depleting highly abundant proteins such as albumin and immunoglobulins to reach out proteins of low abundance and overcome their masking effects. Several commercial kits and methods are routinely practiced during serum/plasma proteomics. Besides plasma and serum other biological fluids have been utilized for biomarker discovery with limited success.

9.5 Potentials Biomarker Candidates in High Altitude Pathophysiology

Several proteomics studies have already been conducted over several decades across various laboratories on humans at high altitudes or in contained experimental hypoxic models using hypobaric chambers. These studies have reported several potential biomarkers based on proteomics studies. Studies by Sharma et al. using 2D gel electrophoresis reported SULT1A1 as a potential biomarker associated with gradual changes in the pulmonary proteome of those exposed to hypobaric hypoxia. These studies were conducted on rats at a simulated altitude of 7600 m, approximating some of the highest mountain peaks in the greater Himalayas (Ahmad et al. 2015). Ahmad and Sharma also performed extensive studies on plasma and serum proteome profiling of rats and humans in independent studies and demonstrated the upregulation of several proteins including vitamin D-binding protein, hemopexin, alpha-1-antitrypsin, haptoglobin β-chain, apolipoprotein A1, transthyretin, and hemoglobin beta chain while downregulation of transferrin, complement C3, serum amyloid, complement component 4A, and plasma retinol-binding protein (Ahmad et al. 2013). In similar studies on rats, hypobaric hypoxia-induced changes in several proteins including Ttr, Prdx-2, Gpx-3, Apo A-I, Hp, Apo-E were recorded (Ahmad et al. 2014). In yet another study by Sharma et al., plasma proteomics of high altitude pulmonary edema (HAPE) patients, a few proteins, namely acute phase proteins (APPs), complement components, and apolipoproteins among others. Among the APPs, haptoglobin α2 chain, haptoglobin β chain, transthyretin, and plasma retinol-binding precursor were found to be differentially expressed indicating their potential for being developed as a biomarker (Ahmad et al. 2011). Brain, which is also the most-affected organ due to hypobaric hypoxia, was also evaluated for proteomics changes in experimental conditions and it was observed that glycolytic enzymes like Gapdh, Pgam1, Eno1, and malate-aspartate shuttle enzymes Mdh1 and Got1in the cortex as compared to hippocampus deciphering efficient use of energy-producing substrates. This was coupled with a concomitant increase in the expression of antioxidant enzymes like Sod1, Sod2, and

Pebp1 in the cortex (Ahmad et al. 2011). In yet another study by Gayatri et al., 2D gel electrophoresis-based proteomic analysis has revealed proteomics markers for hypoxia susceptibility in experimental rats. They reported upregulation of several antioxidant proteins, namely TTR, GPx-3, PON1, Rab-3D, CLC11, CRP, and Hp in hypoxia tolerant rats, while apolipoprotein A-I (APOA1) was upregulated in hypoxia susceptible rats. Furthermore, proteomics analysis of Ladakhi natives using 2DGE followed by MALDI-TOF showed functional regulation between the renin–angiotensin system and eNOS-cGMP pathway and concomitant elevation in the levels of eNOS, phosphorylated eNOS (Ser1177), and plasma biomarkers for nitric oxide (NO) production (nitrate and nitrite) as well as the availability of cGMP (Padhy et al. 2017). 2D electrophoresis is not considered a highly accurate technique at present due to non-reproducibility issues (Magdeldin et al. 2014) and hence establishing the differentially regulated proteins in the aforementioned studies remains a challenge. Furthermore, with the advancement in the proteomics approaches, recent studies were performed using state-of-the-art proteomics approaches such as iTRAQ based quantitative proteomics and tandem mass tag-based quantitative proteomics. More recently in a study by Pooja et al., tandem mass tags (TMT) based proteomics showed elevated plasma concentration of apolipoproteins APOB, APOCI, APOCIII, APOE, and APOL, and carbonic anhydrases (CA1 and CA2) during hypoxia exposure which was also corroborated with lipid profiling suggesting a potential perturbation in lipid transport and lipoprotein-associated metabolic and molecular pathways (Pooja et al. 2021). These studies with advanced proteomic approaches suggest that identification of a single proteomic marker is barely possible for such a complex physiological perturbation and hence a larger focus should lie on understanding and proposing entire biochemical pathways or protein networks. Some of the common biological processes which can be potential hubs of biomarker discovery for hypobaric hypoxia-induced pathology or adaptation are elaborated in the following text.

9.5.1   Antioxidant Signaling

One of the most pertinent hubs of protein interactions observed across the aforementioned proteomic studies was the effect on redox milieu and antioxidant signaling. Several highly interacting proteins such as superoxide dismutase, sulfotransferase, thioredoxin, glutathione peroxidase are often shown to be altered during hypoxic insult in both experimental animals and humans as evident from several past studies. Antioxidant signaling is therefore on the key event in controlling (Fig. 9.1).

9.5.2   Lipid Metabolism

Several proteomic and lipidomic studies conducted on high altitude dwelling individuals or high altitude travelers have indicated a significantly changed lipid

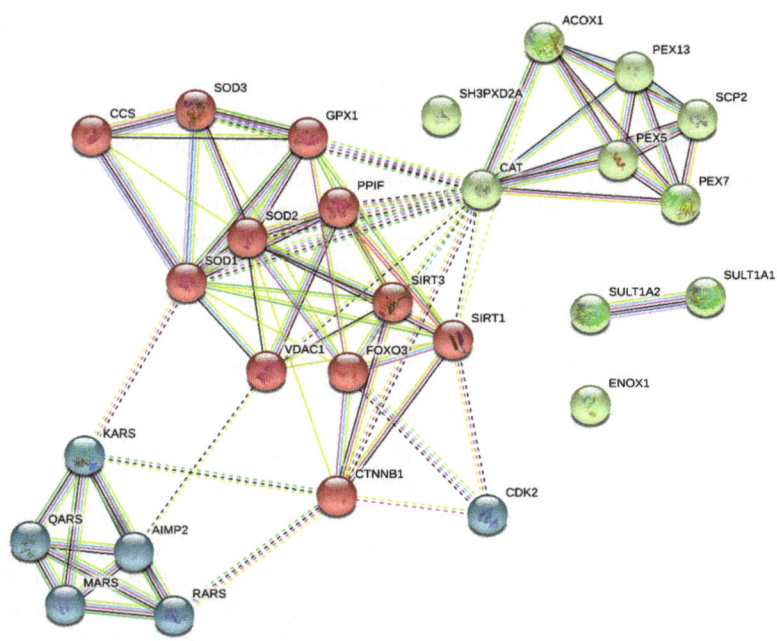

Fig. 9.1 Biological network of potential redox regulating proteins known to show altered expression in hypobaric hypoxia with a potential to be used as biomarkers (K-means clustering, FDR <1%)

profile and lipid regulated pathways perturbations, among which albumin is known to play a nodal role. Albumin, which acts as a carrier of several lipids post-lipogenesis in adipose tissues, is mostly held accountable for hypoxia-induced edema. Concurrently, proteins interacting with albumin are among the potential biomarker candidates which form yet central nodes in hypoxia-perturbed proteomic networks (Fig. 9.1).

9.5.3 Cytoskeleton Remodeling

More recently, studies conducted by Paul et al. observed that an interplay of lung cytoskeletal elements exists that helps in achieving redox homeostasis and extended survival in hypoxic environments. Qualitative perturbations to cytoskeletal stability and innate immunity/inflammation were also observed during extended low pO_2 exposure in humans exposed to 14,000 ft. for 7, 14, and 21 days (Paul et al. 2021). However, to date, scanty information is available for presuming it as a potential biomarker and thus further studies are required to be conducted in a more extended manner to identify potential makers from the cytoskeletal remodeling during hypoxia.

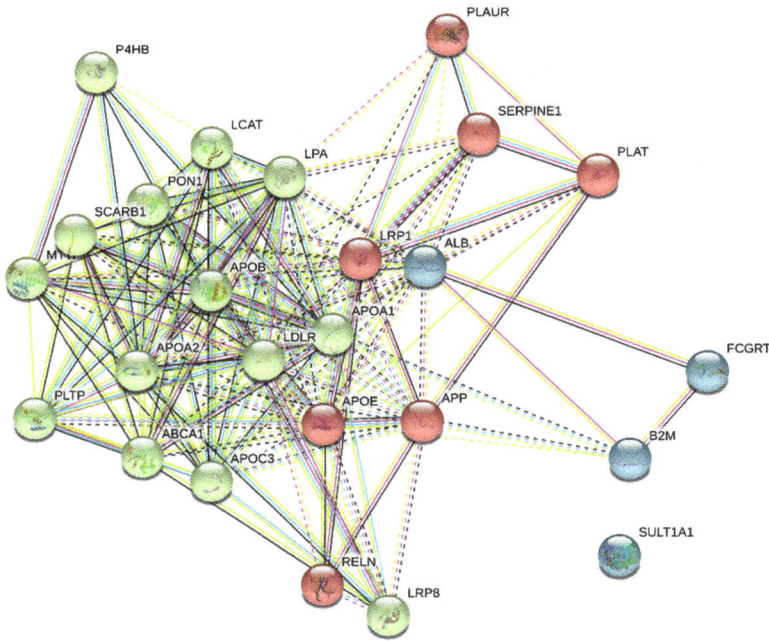

Fig. 9.2 Biological network of potential lipid metabolism-related proteins with albumin occupying a nodal place and hence suggesting a highly orchestrated interplay in hypoxia-mediated changes in lipid metabolism

9.5.4 Post-Translational Modifications

Post-translational modifications induced by hypoxia have recently gained the attention of researchers, especially those mediated by redox signaling. Among the prominent hypoxia-induced protein modifications are carbonylation and nitrosylation. Due to their wide occurrence and diversity in pattern, carbonylation and nitrosylation are possibly the next generation biomarkers for hypoxia-induced stress at the proteomic level. Shotgun and targeted proteomics approaches using biotin switches and DNPH pulldown are often used for the analysis of these modifications. In a study by Anamika et al., the relative carbonylation and nitrosylation in hypoxia-induced samples have been studied with proteome-wide mass spectrometry studies and potential sites of these modifications have been highlighted (Gangwar et al. 2022) and suggested a direct and indirect interaction between nitrosylation and carbonylation pertaining to blood-coagulation and inflammation networks provoked by redox signaling.

9.6   Future Prospects and Conclusion

The future of serum and plasma-based biomarker discovery for high altitude pathology and adaptability is promising. With over several decades of significant scientific contribution and rapid advancements in the proteomic sciences, much clarity on proteomic perturbations has been achieved, yet we are steps away from an exact identification of proteomic markers. In fact, the pathology of high altitude-related changes is so complex that it is very unlikely that a single proteomic marker would be sufficient and achievable for diagnosis or prognosis. However, it is suggested that large-scale proteomic screens or panels with multiplexing features including the expression analysis as well as a post-translational modification would be highly valuable for diagnosis and screening.

References

Ahmad Y, Shukla D, Garg I, Sharma NK, Saxena S, Malhotra VK, Bhargava K (2011) Identification of haptoglobin and apolipoprotein A-I as biomarkers for high altitude pulmonary edema. Funct Integr Genomics 11(3):407–417

Ahmad Y, Sharma NK, Garg I, Ahmad MF, Sharma M, Bhargava K (2013) An insight into the changes in human plasma proteome on adaptation to hypobaric hypoxia. PLoS One 8(7):e67548

Ahmad Y, Sharma NK, Ahmad MF, Sharma M, Garg I, Bhargava K (2014) Proteomic identification of novel differentiation plasma protein markers in hypobaric hypoxia-induced rat model. PLoS One 9(5):e98027

Ahmad Y, Sharma NK, Ahmad MF, Sharma M, Garg I, Srivastava M, Bhargava K (2015) The proteome of hypobaric induced hypoxic lung: insights from temporal proteomic profiling for biomarker discovery. Sci Rep 5:10681

Aslam B, Basit M, Nisar MA, Khurshid M, Rasool MH (2017) Proteomics: technologies and their applications. J Chromatogr Sci 55(2):182–196

Blonder J, Johann DJ, Veenstra TD, Xiao Z, Emmert-Buck MR, Ziegler RG, Rodriguez-Canales J, Hanson JA, Xu X (2008) Quantitation of steroid hormones in thin fresh frozen tissue sections. Anal Chem 80(22):8845–8852

Gallagher S, Chakavarti D (2008) Immunoblot analysis. J Vis Exp 16:759

Gangwar A, Paul S, Ahmad Y, Bhargava K (2020) Intermittent hypoxia modulates redox homeostasis, lipid metabolism associated inflammatory processes and redox post-translational modifications: benefits at high altitude. Sci Rep 10(1):7899

Gangwar A, Paul S, Arya A, Ahmad Y, Bhargava K (2022) Altitude acclimatization via hypoxia-mediated oxidative eustress involves interplay of protein nitrosylation and carbonylation: a redoxomics perspective. Life Sci 296:120021

Gao Z, Luo G, Ni B (2017) Progress in mass spectrometry-based proteomics in hypoxia-related diseases and high-altitude medicine. OMICS 21(6):305–313

Leeman M, Choi J, Hansson S, Storm MU, Nilsson L (2018) Proteins and antibodies in serum, plasma, and whole blood-size characterization using asymmetrical flow field-flow fractionation (AF4). Anal Bioanal Chem 410(20):4867–4873

Lopez LR, Cantella RA, Piscoya Z, Colichon AA, Delgado M, Recavarren S (1975) Immunological survey in high altitude: effect on antibody production and the complement system. Ann Sclavo 17(6):769–785

Magdeldin S, Enany S, Yoshida Y, Xu B, Zhang Y, Zureena Z, Lokamani I, Yaoita E, Yamamoto T (2014) Basics and recent advances of two dimensional-polyacrylamide gel electrophoresis. Clin Proteomics 11(1):16

Matthiesen R, Bunkenborg J (2013) Introduction to mass spectrometry-based proteomics. Methods Mol Biol 1007:1–45

Nedelkov D, Kiernan UA, Niederkofler EE, Tubbs KA, Nelson RW (2005) Investigating diversity in human plasma proteins. Proc Natl Acad Sci U S A 102(31):10852–10857

Padhy G, Gangwar A, Sharma M, Bhargava K, Sethy NK (2017) Plasma proteomics of Ladakhi natives reveal functional regulation between renin-angiotensin system and eNOS-cGMP pathway. High Alt Med Biol 18(1):27–36

Paul S, Gangwar A, Arya A, Bhargava K, Ahmad Y (2021) Modulation of lung cytoskeletal remodeling, RXR based metabolic cascades and inflammation to achieve redox homeostasis during extended exposures to lowered pO2. Apoptosis: Int J Program Cell Death 26(7–8):431–446

Pooja SV, Meena RN, Ray K, Panjwani U, Varshney R, Sethy NK (2021) TMT-based plasma proteomics reveals dyslipidemia among lowlanders during prolonged stay at high altitudes. Front Physiol 12:730601

Veenstra TD (2007) Global and targeted quantitative proteomics for biomarker discovery. J Chromatogr B Analyt Technol Biomed Life Sci 847(1):3–11

Zhu W, Smith JW, Huang CM (2010) Mass spectrometry-based label-free quantitative proteomics. J Biomed Biotechnol 2010:840518

Saliva Proteomics as Non-Invasive Application for Biomarker Studies

10

Shikha Jain, Kalpana Bhargava, and Yasmin Ahmad

Abstract

Nowadays, saliva being a source of broad-spectrum biomolecules (mainly proteins, lipids, hormones, and nucleic acids that originated from various local/ systemic sources) holds a promising future among diagnostic samples. Compared to blood, the use of saliva is advantageous because sample collection and processing are easy, minimally invasive, low cost, and better tolerated by individuals. Saliva proteome analysis can therefore give valuable contributions in understanding the pathophysiology of diseases and provide a foundation for the recognition of potential protein markers. A pathophysiological condition caused by an ascent to a high altitude named hypobaric hypoxia occurs due to deficiency of oxygen at the tissue level caused by lower atmospheric pressure of oxygen. Although few reports have documented the effect of exposure to hypobaric hypoxia on plasma and tissue proteome, salivary proteome-based studies remain uninvestigated. Therefore, identification of molecular signatures having key roles in hypobaric hypoxia by analyzing the salivary proteome founds promising. Through salivary proteome, a few proteins such as alpha-enolase, cystatins, apoptosis inducible factor 2, prolactin inducible protein, carbonic anhydrase 6, phospholipid transfer protein (PLTP), interleukin 1 receptor antagonist (IL1R1), albumin, alpha-1 acid glycoprotein, and alpha-1 antitrypsin were found to be evolved as biomarkers for hypobaric hypoxia. In conclusion, these studies provided the proofs of concept for translating salivary proteins into a non-invasive putative diagnostic panel for assessing hypobaric hypoxia.

S. Jain · K. Bhargava · Y. Ahmad (✉)
Peptide & Proteomics Division, Defence Institute of Physiology & Allied Sciences (DIPAS), Defence Research & Development Organisation (DRDO), Timarpur, New Delhi, Delhi, India

N. K. Sharma, A. Arya (eds.), *High Altitude Sickness – Solutions from Genomics, Proteomics and Antioxidant Interventions*,
https://doi.org/10.1007/978-981-19-1008-1_10

Keywords

Saliva · Hypobaric hypoxia · Proteomics · Biomarker

10.1   Saliva: A Novel Informative Sample

Saliva is an oral fluid that originates from major salivary glands such as parotid, sublingual, and submandibular and minor salivary glands such as labial, buccal, lingual, and palatal glands (Fig. 10.1) (Carranza et al. 2005; Forde et al. 2006; Yoshizawa et al. 2013).

Some of the salivary components may not have originated from salivary glands as saliva also contains fluids from oral mucosal cells, upper respiratory secretions, and gastro-intestinal reflux (Mager et al. 2005; Zhang et al. 2010). Salivary glands are enveloped by capillaries and highly permeable for exchanging blood based molecules into saliva freely (Fig. 10.2) (Haeckel and Hanecke 1996). Also, blood based compounds move from plasma to saliva involving several processes (Yoshizawa et al. 2013; Yamaguchi et al. 2005). These processes include a) ultra-filtration of molecules such as water, ions, catecholamines, and steroids through gap junctions between secretory cells; b) selective transport (actively or passively)

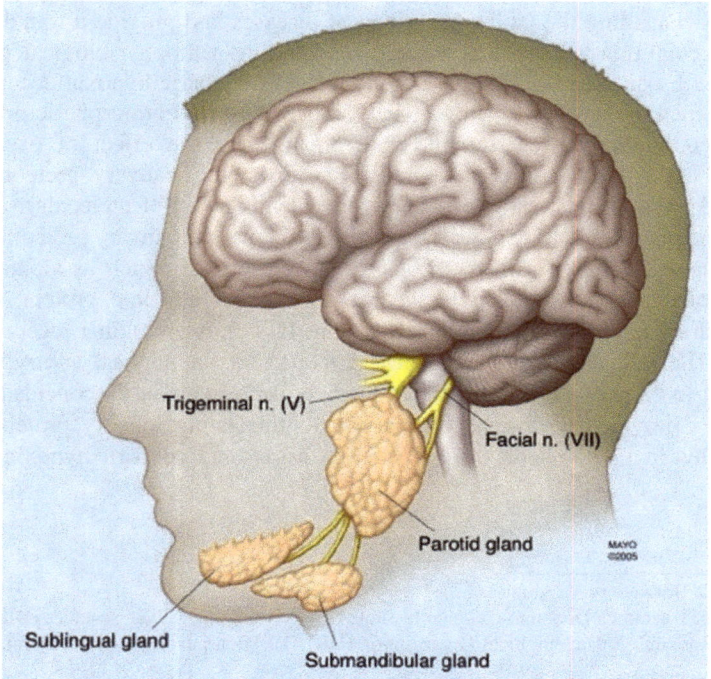

Fig. 10.1 Major salivary glands. (Adapted from Forde et al. 2006)

Fig. 10.2 Mechanism of molecular transport from blood to saliva through capillaries. (Adapted from Haeckel and Hanecke 1996)

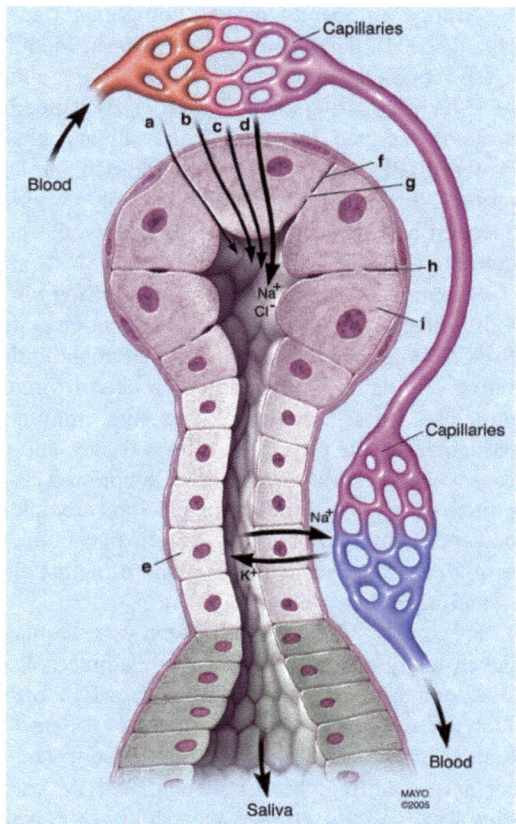

through cellular membranes; and c) transudation of albumin directly from cervical fluid into oral cavity (Chiappin et al. 2007).

Healthy individuals produce 500–1500 mL of saliva (slightly acidic clear fluid having pH = 6.0–7.0) per day (Chicharro et al. 1998; Edgar 1990; Humphrey and Williamson 2001; Van Nieuw Amerongen et al. 2004; Zalewska et al. 2000). Several pathophysiological conditions can modify the amount of saliva (Chicharro et al. 1998; Aps and Martens 2005; Walsh et al. 2004). Saliva contains a variety of compounds such as inorganic (water, ions); organic (non-proteins such as uric acid, bilirubin, creatine, glucose, lipids, amines, and lactate); proteins and hormones (catecholamines and steroids) (Chicharro et al. 1998; Actis et al. 2005; Agha-Hosseini et al. 2006; Cooke et al. 2003; Coufal et al. 2003; Diab-Ladki et al. 2003; Guan et al. 2004; Larsson et al. 1996; Lloyd et al. 1996; Nagler et al. 2002; Rehak et al. 2000; Zelles et al. 1995). These biomolecules being present in the saliva provide information from several organs and systems and raise the possibility of their use as disease biomarkers.

A huge number of protein components are present in saliva (Hu et al. 2005; Huang 2004). These proteins have been identified using proteomics tools such as 2-DE coupled with MALDI-TOF/TOF followed by shotgun proteomics and LC-MS/MS (Castagnola et al. 2017). By using these techniques, most of the salivary proteins secreted in saliva from salivary glands were found to be proline-rich proteins (PRPs), mucins, cysteine-rich proteins (cystatins), and histidine-rich proteins (histatins). PRPs are highly polymorphic in nature with 50 different proteins encoded by different gene arrangements and post-secretory processing of only six genes (Carpenter 2013).

Secretion from each salivary gland differs in concentration of proteins and salts/ions (Hu et al. 2004; Kalk et al. 2002). For example, sublingual gland secretes mucin MUC5B and calgranulin, whereas submandibular gland secretes cystatin C. Human salivary proteins have functions related to immune defense in which a variety of proteins such as lactoferrin, lysozyme, immunoglobulins, agglutinins, and mucins participate in the protection of oral tissues and proteins like histatins and defensins possess bactericidal properties (Chiappin et al. 2007). Other functions of salivary proteins include inhibition of calcium precipitation by PRPs and statherins; taste perception by carbonic anhydrase; digestion of starch by amylase; endonuclease activity by Von Ebner minor gland proteins; and proteinase inhibition by cystatins (Amerongen and Veerman 2002).

Previously, blood/serum/plasma was frequently used as the source of biomarker but in many conditions, the blood sample collection could be problematic, expensive, and invasive. Comparatively, saliva offers various advantages over blood such as: (a) quick and easy method of sample collection: non-requirement of a trained professional and self-collection; (b) non-invasive: painless collection; (c) safe handling: diseases such as HIV cannot be transmitted through saliva samples; (d) easy storage and shipping as saliva does not clot like blood; and (e) cost-effective (Yoshizawa et al. 2013; Chiappin et al. 2007; Schafer et al. 2014; Kaczor-Urbanowicz et al. 2017; Lee and Wong 2009; Pfaffe et al. 2011; Campo et al. 2006). Recently, it gained attention as an effective strategy for screening, diagnosis, prognosis, and monitoring post-therapy status. And, it is imperative to explore potential of saliva based diagnostics as discussed in the following section.

10.2   Hypobaric Hypoxia: A Pathophysiological Condition

Ascent of an individual to high altitude, the parallel low atmospheric pressure of oxygen leads to decreased body's partial pressure of oxygen, thereby causing a pathophysiological condition named hypobaric hypoxia (HH) (Fig. 10.3). HH affects the ability of body to exchange oxygen between the lungs and the bloodstream, thereby disrupting the oxygen availability to the tissues. A significant population of world resides at an elevation of 10,000 ft. above sea level. This population consists of people who have lived there for generations and travelers who travel to high altitude intermittently (soldiers) or for short to moderate periods

Fig. 10.3   Oxygen levels at different altitude elevations. (Adapted from http://www.altitude.org)

of time (e.g. mountain climbers) (Askew 2002). The people residing at high altitude for generations have already evolved various mechanisms to generate energy at high altitude (Hoppeler and Vogt 2001). Whereas travelers may experience AMS or potentially more serious impairments such as HACE and HAPE (Hackett and Roach 2001, 2004; San et al. 2013; Sharp and Bernaudin 2004; Wilson et al. 2009). These maladies develop over hours to days at high altitude and are known to be preventable but still remain common consequences of rapid ascent to high altitude.

10.3   Redox Stress: Molecular Responses in Hypobaric Hypoxia

The main consequence of hypobaric hypoxia is generation of reactive oxygen and nitrogen species (RONS). RONS, physiological modulators of cellular redox mechanism control wide range of physiological and pathophysiological processes (Bakonyi and Radak 2004). Hypoxia is caused by the limited availability of oxygen in mitochondria for reduction to H_2O at cytochrome oxidase and leads to generation of RONS. In order to produce energy (ATP), auto-oxidation of one or more mitochondrial complexes, such as the ubiquinone–ubiquinol redox couple results in accumulation of reducing equivalents (RONS) (Fig. 10.4). Despite the presence of antioxidant system, the levels of RONS generation can lead to oxidative stress (Askew 2002). These accumulated oxidants such as RONS cause imbalance between oxidative stressors and antioxidant capacity. Oxidants and antioxidants

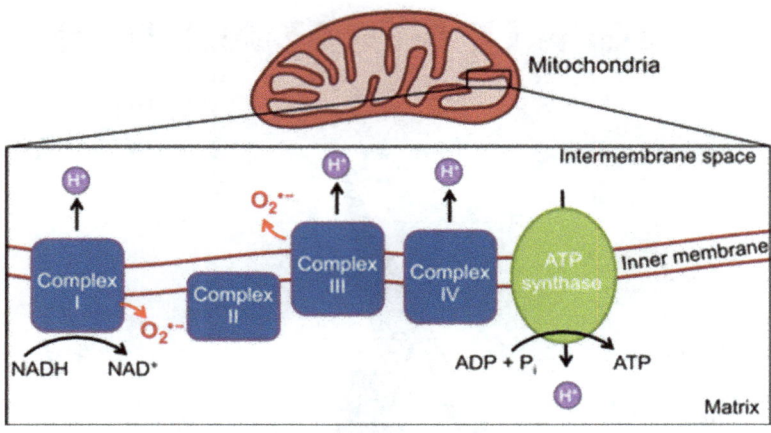

Fig. 10.4 Oxidants generation in mitochondria. (Adapted from Bigarella et al. 2014)

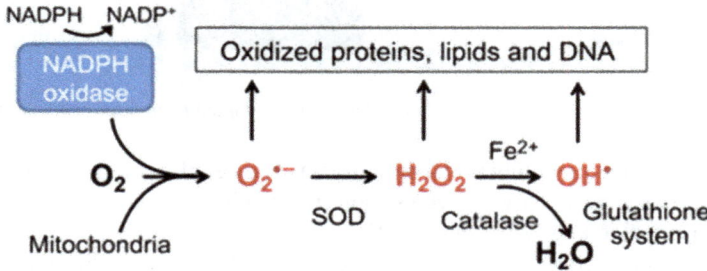

Fig. 10.5 Oxidative stress to biomolecules. (Adapted from Bigarella et al. 2014)

imbalance leads to damage in biomolecules such as lipids, proteins, and DNA (Fig. 10.5) (Dosek et al. 2007; Jefferson et al. 2004; Strapazzon et al. 2016).

This oxidative damage to biomolecules is nearly similar to the stress associated with cancer, neurological, pulmonary, and cardiovascular diseases (Reuter et al. 2010).

In literature, the increased production of oxidants has been documented in blood, tissue, urine, and breath samples of rats and humans in response to hypoxia (Strapazzon et al. 2016; Chang et al. 1989; Irarrazaval et al. 2017; Magalhaes et al. 1985; Maiti et al. 2006; Radak et al. 1994; Ribon et al. 2016; Siervo et al. 2014; Singh et al. 2013; Yoshikawa et al. 1982).

In vivo, aerobic cells develop antioxidant system to regulate the effects of RONS. This system contains mitochondrial, cytosolic, and extra-cellular superoxide dismutase (SOD). SOD converts reactive superoxide ions to H_2O_2 which is then decomposed into water by the action of glutathione system and catalase. Previously, high altitude related studies measured the content of glutathione, SOD, and GPX (Askew 2002; Bakonyi and Radak 2004; Radak et al. 1994, 1997; Nakanishi et al.

1995; Imai et al. 1995; Ilavazhagan et al. 2001; Joanny et al. 2001). And, it appeared that the capacity of antioxidant system was found to be decreasing at high altitude. These changes in redox parameters are responsible behind physiological and patho-physiological changes in the body.

10.4 Use of Biological Fluids Such as Plasma/Serum in Search of Potential Protein Markers

In hypobaric hypoxia related molecular studies, particularly proteomics, biological fluids such as blood/plasma/serum have been used among clinicians and biological scientists in search of potential protein markers for hypobaric hypoxia and acclimatization. Multiple scattered plasma proteins have been identified in samples from patients suffering from high altitude illnesses such as AMS, HAPE, and HACE (Lu et al. 2018; van Patot et al. 2005; Droma et al. 1996; Sartori et al. 1999; Charu et al. 2006; Berger et al. 2009; Barker et al. 2016; Ahmad et al. 2011, 2015; Zhang et al. 2013; Sikri and Bhattachar 2017).

Another important aspect is that the redox imbalance could also trigger healthy cellular processes that aid in acclimatization to hypobaric hypoxia. Protein markers have been studied in rats and humans to understand such cellular processes in acclimatized state (Ahmad et al. 2013, 2014, 2015; Siervo et al. 2014; Luks et al. 2017; Levett et al. 2011; Hartmann et al. 2000).

Newer strategies that are completely non-invasive in nature such as study of saliva have been implicated for diagnostics purposes. Thus, we move on to newer studies for exploration and improvement.

10.5 Saliva as a Diagnostic Fluid in Translational Studies

Previous findings have described that chronic diseases such as cancer, cardiovascular, diabetes, pulmonary, and neurological diseases are associated with continued oxidative stress (Reuter et al. 2010). In translational research, saliva has been explored for the detection of oral cancer (de Jong et al. 2010; Hu et al. 2007a; Gallo et al. 2016), Sjogren's syndrome (Giusti et al. 2007; Hu et al. 2007b; Peluso et al. 2007), breast cancer (Streckfus et al. 2008), lung cancer (Xiao et al. 2012), and systemic disorders such as hepatitis, HIV, and HCV (Elsana et al. 1998; Yaari et al. 2006; Hodinka et al. 1998). A previous study by Shen Hu et al. provided a proof for exploring salivary proteins in oral cancer and revealed thioredoxin as a salivary biomarker for human oral cancer (Hu et al. 2007a). Other researchers, Ebbing P. de Jong et al. reported myosin and actin as promising salivary biomarkers for distinguishing between pre-malignant and malignant oral lesions (de Jong et al. 2010). Another recent study done by Eva Csosz et al. highlighted the importance of identification of population tailored biomarkers and reported oral squamous cell carcinoma (OSCC) biomarkers in a Hungarian population. And, S100A9 and IL-6 were shown to be candidate biomarkers for OSCC (Csosz et al. 2017). In a study performed by G. Peluso et al. in Sjogren's syndrome (SS), saliva collected from patients with primary SS revealed higher levels of alpha-defensin 1 and the presence of beta-defensin 2 which could be potential markers of oral inflammation in SS patients group (Peluso et al. 2007). Omer Deutsch et al. observed profilin and CA-I as biomarker candidates for Sjögren's syndrome following high-abundance protein depletion (Deutsch et al. 2015). In HIV and HCV, rapid point-of-care HIV tests utilize oral fluids to rapidly provide test results to patients (Hodinka et al. 1998; Fernandez Rodriguez et al. 1994). In cases of non-oral cancers such as lung and breast cancers, various researchers suggested modifications in the salivary proteome and provided proof for candidate biomarkers (Streckfus et al. 2008; Xiao et al. 2012; Bigler et al. 2009; Streckfus and Bigler 2016). Another example of its use for determining hormone levels, including estradiol, progesterone and testosterone, DHEA, and cortisol (Groschl 2008).

10.6   Saliva in Response to Hypobaric Hypoxia

In the context of events related to hypobaric hypoxia, studies on saliva have been performed due to its diagnostic potential. An initial study, observed in 1990s, suggested an increase in salivary flow rate and low concentration of potassium in response to acute hypobaric hypoxia exposure (Pilardeau et al. 1990). Another researcher, McLean reported decreased levels of aldosterone in response to both ACTH and renin–angiotensin stimulation at high altitude (McLean et al. 1989). Additionally, a rise in the salivary activities of aminotransferases was observed while HH exposure (Mominzadeh et al. 2014). A recent study by Woods DR et al. reported an alteration in the levels of salivary cortisol and suggested that an elevated cortisol levels may contribute to fluid retention associated with acute mountain sickness (Woods et al. 2012).

In molecular omics-based studies, preferably proteomics, only limited studies have been performed so far. Jain et al. have reported significantly altered proteins such as alpha-enolase, cystatin SN, apoptosis inducing factor 2 (AIF-2), cystatin S, carbonic anhydrase 6 (CA6), and prolactin inducible protein (PIP) and plausible pathways involving these proteins such as inflammation, impaired glycolysis, and respiratory alkalosis during HH exposure (Jain et al. 2018). Another study by Jain et al. suggested proteins such as albumin, carbonic anhydrase 6, prolactin inducible protein, alpha-enolase, phospholipid transfer protein (PLTP), alpha-1 acid glycoprotein, interleukin 1 receptor antagonist (IL1RA), and alpha-1 antitrypsin as protein candidates for assessing hypobaric hypoxia (Jain et al. 2020).

10.7   Proteins Evolved as Biomarkers for Hypobaric Hypoxia

10.7.1 Alpha-Enolase

Alpha-enolase is a multifunctional enzyme known to be involved in various processes such as allergic responses, growth control as well as glycolysis other than inflammatory hypoxic tolerance. Earlier studies suggested in hypoxic situations, expression of alpha-enolase got differentially modulated and provides protection to cells so as to acclimatize to low oxygen levels through increased anaerobic metabolism (Aaronson et al. 1995; Sedoris et al. 2010). Alpha-enolase has several interacting partners such as glyceraldehyde 3-phosphate dehydrogenase, phosphoglycerate kinase, and pyruvate kinase. Alpha-enolase has earlier been observed in various studies and well-established with hypoxia induced physiological changes (Kim et al. 2006; Semenza et al. 1996), suggesting its role as potential protein marker if carefully evaluated and utilized (Mikuriya et al. 2007; Xu et al. 2005).

10.7.2 Cystatins

Salivary cystatins such as cystatin S and SN were observed in whole human saliva. Cystatin S was known to be present in three forms: mono-phosphorylated, di-phosphorylated, and non-phosphorylated (Isemura et al. 1991) and involved in the mineral balance of the tooth. And cystatin-SN was found to inhibit the human lysosomalcathepsins B, H, and L in vitro (Baron et al. 1999). Recent studies reported salivary cystatins as prospective biomarkers of oral diseases and diabetes (Rudney et al. 2009; Bencharit et al. 2013). Cystatins are not elevated in normal conditions.

10.7.3 Apoptosis Inducing Factor 2 (AIF2)

AIF 2, an oxidoreductase acts as a caspase-independent mitochondrial effector of apoptotic cell death and plays a role in mediating apoptosis response. AIF 2 was over-expressed in response to hypobaric hypoxia, thereby confirming its induction by hypoxia (Greijer and van der Wall 2004).

10.7.4 Prolactin Inducible Protein (PIP)

PIP, an extra-parotid glycoprotein associated with secretory cell differentiation was found to be decreased in hypobaric hypoxia. Earlier, it was observed that decreased levels of salivary PIP were found in patients with bleeding oral cavities (Huang 2004).

10.7.5 Carbonic Anhydrase 6 (CA6)

CA6, an enzyme involved in respiratory alkalosis, could be considered as potential marker (Taylor 2011). As per earlier reports, buffering capacity of saliva plays an important in oral homeostasis and the bicarbonate buffer is the main buffer that contributes to the salivary buffering capacity (Bardow et al. 2000; Peres et al. 2010). Other findings suggested an ascent to high altitude was known to be associated with decreased concentration of bicarbonate and hydrogen ions, increased pH and thus, resulting in respiratory alkalosis (Goldfarb-Rumyantzev and Alper 2014). The increased expression of carbonic anhydrase 6 in this study suggested its probable role in conferring acclimatization to hypobaric hypoxia by neutralizing pH through bicarbonate balance.

10.7.6 Phospholipid Transfer Protein (PLTP)

PLTP, a regulator of lipid metabolism is known to play an important role in oxidative stress. Recent research suggested that PLTP modulates BBB integrity, possibly through its ability to transfer vitamin E, and modulate cerebro-vascular oxidative stress (Zhou et al. 2014). Another study suggested regulation of vitamin E bioavailability in lipoproteins through PLTP in order to protect circulating lipoproteins from oxidative stress (Jiang et al. 2002). Also, a finding suggested that this protein showed enhanced expression in hypoxia stimulus in case of emphysema (Jiang et al. 1998).

10.7.7 Interleukin 1 Receptor Antagonist (IL1R1)

Interleukin 1 receptor antagonist (IL1R1) is an upstream regulator of the pathway and a cytokine widely known to be associated with inflammation and expressed by activated mononuclear cells. The levels of circulating IL1RA in the plasma were increased in response to hypoxia suggesting the reason for its decreased level in salivary secretions and as a potential target for anti-inflammatory therapy (Fritzsching et al. 2015).

10.7.8 Albumin, Alpha-1 Acid Glycoprotein, and Alpha-1 Antitrypsin

These downstream regulators of acute phase response signaling also showed similar patterns as IL1R1. According to previous research, albumin was observed to suppress VEGF via alteration of HIF/HRE pathway (Katavetin et al. 2008). Albumin also provides protection against hypoxia induced injuries (Strubelt et al. 1994). Another protein, alpha-1 acid glycoprotein is known to provide protection against hypoxia by inhibiting inflammation and apoptosis, thereby could be given exogenously as therapeutics (de Vries et al. 2004; Hochepied et al. 2003; Van Molle et al.

1997). Similar to other downstream regulators of the pathway, alpha-1 antitrypsin also has anti-inflammatory properties (Ahmad et al. 2013).

10.8 Concluding Remarks

In conclusion, these studies provided the proof of concept for translating salivary proteins into a non-invasive putative diagnostic panel for assessing hypobaric hypoxia. This panel has the potential to be used in future to diagnose individuals affected by hypobaric hypoxia.

References

Aaronson RM, Graven KK, Tucci M, McDonald RJ, Farber HW (1995) Non-neuronal enolase is an endothelial hypoxic stress protein. J Biol Chem 270:27752–27757

Actis AB, Perovic NR, Defago D, Beccacece C, Eynard AR (2005) Fatty acid profile of human saliva: a possible indicator of dietary fat intake. Arch Oral Biol 50:1–6

Agha-Hosseini F, Dizgah IM, Amirkhani S (2006) The composition of unstimulated whole saliva of healthy dental students. J Contemp Dent Pract 7:104–111

Ahmad Y, Shukla D, Garg I, Sharma NK, Saxena S, Malhotra V, Bhargava K (2011) Identification of haptoglobin and apolipoprotein AI as biomarkers for high altitude pulmonary edema. Funct Integr Genomics 11:407

Ahmad Y, Sharma NK, Garg I, Ahmad MF, Sharma M, Bhargava K (2013) An insight into the changes in human plasma proteome on adaptation to hypobaric hypoxia. PLoS One 8:e67548

Ahmad Y, Sharma NK, Ahmad MF, Sharma M, Garg I, Bhargava K (2014) Proteomic identification of novel differentiation plasma protein markers in hypobaric hypoxia-induced rat model. PLoS One 9:e98027

Ahmad Y, Sharma NK, Ahmad MF, Sharma M, Garg I, Srivastava M, Bhargava K (2015) The proteome of hypobaric induced hypoxic lung: insights from temporal proteomic profiling for biomarker discovery. Sci Rep 5:10681

Amerongen AV, Veerman EC (2002) Saliva—the defender of the oral cavity. Oral Dis 8:12–22

Aps JK, Martens LC (2005) Review: the physiology of saliva and transfer of drugs into saliva. Forensic Sci Int 150:119–131

Askew EW (2002) Work at high altitude and oxidative stress: antioxidant nutrients. Toxicology 180:107–119

Bakonyi T, Radak Z (2004) High altitude and free radicals. J Sports Sci Med 3:64–69

Bardow A, Moe D, Nyvad B, Nauntofte B (2000) The buffer capacity and buffer systems of human whole saliva measured without loss of CO2. Arch Oral Biol 45(1):1–12. https://doi.org/10.1016/s0003-9969(99)00119-3. PMID: 10669087

Barker KR, Conroy AL, Hawkes M, Murphy H, Pandey P, Kain KC (2016) Biomarkers of hypoxia, endothelial and circulatory dysfunction among climbers in Nepal with AMS and HAPE: a prospective case–control study. J Travel Med 23(3):taw005

Baron A, DeCarlo A, Featherstone J (1999) Functional aspects of the human salivary cystatins in the oral environment. Oral Dis 5:234–240

Bencharit S, Baxter SS, Carlson J, Byrd WC, Mayo MV, Border MB, Kohltfarber H, Urrutia E, Howard-Williams EL, Offenbacher S, Wu MC, Buse JB (2013) Salivary proteins associated with hyperglycemia in diabetes: a proteomic analysis. Mol Biosyst 9:2785–2797

Berger MM, Dehnert C, Bailey DM, Luks AM, Menold E, Castell C, Schendler G, Faoro V, Mairbäurl H, Bärtsch P (2009) Transpulmonary plasma ET-1 and nitrite differences in high altitude pulmonary hypertension. High Alt Med Biol 10:17–24

Bigarella CL, Liang R, Ghaffari S (2014) Stem cells and the impact of ROS signaling. Development 141:4206–4218

Bigler LR, Streckfus CF, Dubinsky WP (2009) Salivary biomarkers for the detection of malignant tumors that are remote from the oral cavity. Clin Lab Med 29:71–85

Campo J, Perea MA, del Romero J, Cano J, Hernando V, Bascones A (2006) Oral transmission of HIV, reality or fiction? An update. Oral Dis 12:219–228

Carpenter GH (2013) The secretion, components, and properties of saliva. Annu Rev Food Sci Technol 4:267–276

Carranza M, Ferraris ME, Galizzi M (2005) Structural and morphometrical study in glandular parenchyma from alcoholic sialosis. J Oral Pathol Med 34:374–379

Castagnola M, Scarano E, Passali GC, Messana I, Cabras T, Iavarone F, Di Cintio G, Fiorita A, De Corso E, Paludetti G (2017) Salivary biomarkers and proteomics: future diagnostic and clinical utilities. Acta Otorhinolaryngol Ital 37:94–101

Chang SW, Stelzner TJ, Weil JV, Voelkel NF (1989) Hypoxia increases plasma glutathione disulfide in rats. Lung 167:269–276

Charu R, Stobdan T, Ram R, Khan A, Pasha MQ, Norboo T, Afrin F (2006) Susceptibility to high altitude pulmonary oedema: role of ACE and ET-1 polymorphisms. Thorax 61:1011–1012

Chiappin S, Antonelli G, Gatti R, De Palo EF (2007) Saliva specimen: a new laboratory tool for diagnostic and basic investigation. Clin Chim Acta 383:30–40

Chicharro JL, Lucia A, Perez M, Vaquero AF, Urena R (1998) Saliva composition and exercise. Sports Med 26:17–27

Cooke M, Leeves N, White C (2003) Time profile of putrescine, cadaverine, indole and skatole in human saliva. Arch Oral Biol 48:323–327

Coufal P, Zuska J, van de Goor T, Smith V, Gas B (2003) Separation of twenty underivatized essential amino acids by capillary zone electrophoresis with contactless conductivity detection. Electrophoresis 24:671–677

Csosz E, Labiscsak P, Kallo G, Markus B, Emri M, Szabo A, Tar I, Tozser J, Kiss C, Marton I (2017) Proteomics investigation of OSCC-specific salivary biomarkers in a Hungarian population highlights the importance of identification of population-tailored biomarkers. PLoS One 12: e0177282

Deutsch O, Krief G, Konttinen YT, Zaks B, Wong DT, Aframian DJ, Palmon A (2015) Identification of Sjogren's syndrome oral fluid biomarker candidates following high-abundance protein depletion. Rheumatology (Oxford) 54:884–890

Diab-Ladki R, Pellat B, Chahine R (2003) Decrease in the total antioxidant activity of saliva in patients with periodontal diseases. Clin Oral Investig 7:103–107

Dosek A, Ohno H, Acs Z, Taylor AW, Radak Z (2007) High altitude and oxidative stress. Respir Physiol Neurobiol 158:128–131

Droma Y, Hayano T, Takabayashi Y, Koizumi T, Kubo K, Kobayashi T, Sekiguchi M (1996) Endothelin-1 and interleukin-8 in high altitude pulmonary oedema. Eur Respir J 9:1947–1949

Edgar WM (1990) Saliva and dental health. Clinical implications of saliva: report of a consensus meeting. Br Dent J 169:96–98

Elsana S, Sikuler E, Yaari A, Shemer-Avni Y, Abu-Shakra M, Buskila D, Katzman P, Naggan L, Margalith M (1998) HCV antibodies in saliva and urine. J Med Virol 55:24–27

Fernandez Rodriguez E, Carcaba Fernandez V, Rodriguez Junquera M, Alfonso Megido J, Garcia Amorin Z, Garcia Alonso S (1994) Detection of HIV antibodies in saliva using a rapid diagnostic immunoenzyme assay. Rev Clin Esp 194:523–525

Forde MD, Koka S, Eckert SE, Carr AB, Wong DT (2006) Systemic assessments utilizing saliva: Part 1 general considerations and current assessments. Int J Prosthodont 19:43–52

Fritzsching B, Zhou-Suckow Z, Trojanek JB, Schubert SC, Schatterny J, Hirtz S, Agrawal R, Muley T, Kahn N, Sticht C, Gunkel N, Welte T, Randell SH, Langer F, Schnabel P, Herth FJ, Mall MA (2015) Hypoxic epithelial necrosis triggers neutrophilic inflammation via IL-1 receptor signaling in cystic fibrosis lung disease. Am J Respir Crit Care Med 191:902–913

Gallo C, Ciavarella D, Santarelli A, Ranieri E, Colella G, Lo Muzio L, Lo Russo L (2016) Potential salivary proteomic markers of Oral squamous cell carcinoma. Cancer Genomics Proteomics 13: 55–61

Giusti L, Baldini C, Bazzichi L, Ciregia F, Tonazzini I, Mascia G, Giannaccini G, Bombardieri S, Lucacchini A (2007) Proteome analysis of whole saliva: a new tool for rheumatic diseases—the example of Sjogren's syndrome. Proteomics 7:1634–1643

Goldfarb-Rumyantzev AS, Alper SL (2014) Short-term responses of the kidney to high altitude in mountain climbers. Nephrol Dial Transplant 29(3):497–506. https://doi.org/10.1093/ndt/gft051. Epub 2013 Mar 22. PMID: 23525530; PMCID: PMC3938295

Greijer AE, van der Wall E (2004) The role of hypoxia inducible factor 1 (HIF-1) in hypoxia induced apoptosis. J Clin Pathol 57:1009–1014

Groschl M (2008) Current status of salivary hormone analysis. Clin Chem 54:1759–1769

Guan Y, Chu Q, Ye J (2004) Determination of uric acid in human saliva by capillary electrophoresis with electrochemical detection: potential application in fast diagnosis of gout. Anal Bioanal Chem 380:913–917

Hackett PH, Roach RC (2001) High-altitude illness. N Engl J Med 345:107–114

Hackett PH, Roach RC (2004) High altitude cerebral edema. High Alt Med Biol 5:136–146

Haeckel R, Hanecke P (1996) Application of saliva for drug monitoring. An in vivo model for transmembrane transport. Eur J Clin Chem Clin Biochem 34:171–191

Hartmann G, Tschöp M, Fischer R, Bidlingmaier C, Riepl R, Tschöp K, Hautmann H, Endres S, Toepfer M (2000) High altitude increases circulating interleukin-6, interleukin-1 receptor antagonist and C-reactive protein. Cytokine 12:246–252

Hochepied T, Berger FG, Baumann H, Libert C (2003) Alpha(1)-acid glycoprotein: an acute phase protein with inflammatory and immunomodulating properties. Cytokine Growth Factor Rev 14: 25–34

Hodinka RL, Nagashunmugam T, Malamud D (1998) Detection of human immunodeficiency virus antibodies in oral fluids. Clin Diagn Lab Immunol 5:419–426

Hoppeler H, Vogt M (2001) Muscle tissue adaptations to hypoxia. J Exp Biol 204:3133–3139

Hu S, Denny P, Denny P, Xie Y, Loo JA, Wolinsky LE, Li Y, McBride J, Ogorzalek Loo RR, Navazesh M, Wong DT (2004) Differentially expressed protein markers in human submandibular and sublingual secretions. Int J Oncol 25:1423–1430

Hu S, Xie Y, Ramachandran P, Ogorzalek Loo RR, Li Y, Loo JA, Wong DT (2005) Large-scale identification of proteins in human salivary proteome by liquid chromatography/mass spectrometry and two-dimensional gel electrophoresis-mass spectrometry. Proteomics 5:1714–1728

Hu S, Yu T, Xie Y, Yang Y, Li Y, Zhou X, Tsung S, Loo RR, Loo JR, Wong DT (2007a) Discovery of oral fluid biomarkers for human oral cancer by mass spectrometry. Cancer Genomics Proteomics 4:55–64

Hu S, Wang J, Meijer J, Ieong S, Xie Y, Yu T, Zhou H, Henry S, Vissink A, Pijpe J, Kallenberg C, Elashoff D, Loo JA, Wong DT (2007b) Salivary proteomic and genomic biomarkers for primary Sjogren's syndrome. Arthritis Rheum 56:3588–3600

Huang CM (2004) Comparative proteomic analysis of human whole saliva. Arch Oral Biol 49:951–962

Humphrey SP, Williamson RT (2001) A review of saliva: normal composition, flow, and function. J Prosthet Dent 85:162–169

Ilavazhagan G, Bansal A, Prasad D, Thomas P, Sharma SK, Kain AK, Kumar D, Selvamurthy W (2001) Effect of vitamin E supplementation on hypoxia-induced oxidative damage in male albino rats. Aviat Space Environ Med 72:899–903

Imai H, Kashiwazaki H, Suzuki T, Kabuto M, Himeno S, Watanabe C, Moji K, Kim SW, Rivera JO, Takemoto T (1995) Selenium levels and glutathione peroxidase activities in blood in an Andean high-altitude population. J Nutr Sci Vitaminol (Tokyo) 41:349–361

Irarrazaval S, Allard C, Campodonico J, Perez D, Strobel P, Vasquez L, Urquiaga I, Echeverria G, Leighton F (2017) Oxidative stress in acute hypobaric hypoxia. High Alt Med Biol 18:128–134

Isemura S, Saitoh E, Sanada K, Minakata K (1991) Identification of full-sized forms of salivary (S-type) cystatins (cystatin SN, cystatin SA, cystatin S, and two phosphorylated forms of cystatin S) in human whole saliva and determination of phosphorylation sites of cystatin S. J Biochem 110:648–654

Jain S, Ahmad Y, Bhargava K (2018) Salivary proteome patterns of individuals exposed to high altitude. Arch Oral Biol 96:104–112

Jain S, Paul S, Meena RN, Gangwar A, Panjwani U, Ahmad Y, Bhargava K (2020) Saliva panel of protein candidates: a comprehensive study for assessing high altitude acclimatization. Nitric Oxide 95:1–11

Jefferson JA, Simoni J, Escudero E, Hurtado ME, Swenson ER, Wesson DE, Schreiner GF, Schoene RB, Johnson RJ, Hurtado A (2004) Increased oxidative stress following acute and chronic high altitude exposure. High Alt Med Biol 5:61–69

Jiang XC, D'Armiento J, Mallampalli RK, Mar J, Yan SF, Lin M (1998) Expression of plasma phospholipid transfer protein mRNA in normal and emphysematous lungs and regulation by hypoxia. J Biol Chem 273:15714–15718

Jiang XC, Tall AR, Qin S, Lin M, Schneider M, Lalanne F, Deckert V, Desrumaux C, Athias A, Witztum JL, Lagrost L (2002) Phospholipid transfer protein deficiency protects circulating lipoproteins from oxidation due to the enhanced accumulation of vitamin E. J Biol Chem 277:31850–31856

Joanny P, Steinberg J, Robach P, Richalet JP, Gortan C, Gardette B, Jammes Y (2001) Operation Everest III (Comex'97): the effect of simulated sever hypobaric hypoxia on lipid peroxidation and antioxidant defence systems in human blood at rest and after maximal exercise. Resuscitation 49:307–314

de Jong EP, Xie H, Onsongo G, Stone MD, Chen XB, Kooren JA, Refsland EW, Griffin RJ, Ondrey FG, Wu B, Le CT, Rhodus NL, Carlis JV, Griffin TJ (2010) Quantitative proteomics reveals myosin and actin as promising saliva biomarkers for distinguishing pre-malignant and malignant oral lesions. PLoS One 5:e11148

Kaczor-Urbanowicz KE, Martin Carreras-Presas C, Aro K, Tu M, Garcia-Godoy F, Wong DT (2017) Saliva diagnostics—current views and directions. Exp Biol Med (Maywood) 242:459–472

Kalk WW, Vissink A, Stegenga B, Bootsma H, Nieuw Amerongen AV, Kallenberg CG (2002) Sialometry and sialochemistry: a non-invasive approach for diagnosing Sjogren's syndrome. Ann Rheum Dis 61:137–144

Katavetin P, Inagi R, Miyata T, Tanaka T, Sassa R, Ingelfinger JR, Fujita T, Nangaku M (2008) Albumin suppresses vascular endothelial growth factor via alteration of hypoxia-inducible factor/hypoxia-responsive element pathway. Biochem Biophys Res Commun 367:305–310

Kim JW, Tchernyshyov I, Semenza GL, Dang CV (2006) HIF-1-mediated expression of pyruvate dehydrogenase kinase: a metabolic switch required for cellular adaptation to hypoxia. Cell Metab 3:177–185

Larsson B, Olivecrona G, Ericson T (1996) Lipids in human saliva. Arch Oral Biol 41:105–110

Lee YH, Wong DT (2009) Saliva: an emerging biofluid for early detection of diseases. Am J Dent 22:241–248

Levett DZ, Fernandez BO, Riley HL, Martin DS, Mitchell K, Leckstrom CA, Ince C, Whipp BJ, Mythen MG, Montgomery HE, Grocott MP, Feelisch M, For the Caudwell Extreme Everest Research, G (2011) The role of nitrogen oxides in human adaptation to hypoxia. Sci Rep 1:109

Lloyd JE, Broughton A, Selby C (1996) Salivary creatinine assays as a potential screen for renal disease. Ann Clin Biochem 33(Pt 5):428–431

Lu H, Wang R, Li W, Xie H, Wang C, Hao Y, Sun Y, Jia Z (2018) Plasma proteomic study of acute mountain sickness susceptible and resistant individuals. Sci Rep 8:1265

Luks AM, Levett D, Martin DS, Goss CH, Mitchell K, Fernandez BO, Feelisch M, Grocott MP, Swenson ER, the Caudwell Xtreme Everest, I. (2017) Changes in acute pulmonary vascular responsiveness to hypoxia during a progressive ascent to high altitude (5300 m). Exp Physiol 102:711–724

Magalhaes J, Ascensao A, Soares JM, Ferreira R, Neuparth MJ, Marques F, Duarte JA (1985) (2005) acute and severe hypobaric hypoxia increases oxidative stress and impairs mitochondrial function in mouse skeletal muscle. J Appl Physiol 99:1247–1253

Mager DL, Haffajee AD, Devlin PM, Norris CM, Posner MR, Goodson JM (2005) The salivary microbiota as a diagnostic indicator of oral cancer: a descriptive, non-randomized study of cancer-free and oral squamous cell carcinoma subjects. J Transl Med 3:27

Maiti P, Singh SB, Sharma AK, Muthuraju S, Banerjee PK, Ilavazhagan G (2006) Hypobaric hypoxia induces oxidative stress in rat brain. Neurochem Int 49:709–716

McLean CJ, Booth CW, Tattersall T, Few JD (1989) The effect of high altitude on saliva aldosterone and glucocorticoid concentrations. Eur J Appl Physiol Occup Physiol 58:341–347

Mikuriya K, Kuramitsu Y, Ryozawa S, Fujimoto M, Mori S, Oka M, Hamano K, Okita K, Sakaida I, Nakamura K (2007) Expression of glycolytic enzymes is increased in pancreatic cancerous tissues as evidenced by proteomic profiling by two-dimensional electrophoresis and liquid chromatography-mass spectrometry/mass spectrometry. Int J Oncol 30:849–855

Mominzadeh M, Mirzaii-Dizgah I, Mirzaii-Dizgah MR, Mirzaii-Dizgah MH (2014) Stimulated saliva aminotransaminase alteration after experiencing acute hypoxia training. Air Med J 33: 157–160

Nagler RM, Hershkovich O, Lischinsky S, Diamond E, Reznick AZ (2002) Saliva analysis in the clinical setting: revisiting an underused diagnostic tool. J Invest Med 50:214–225

Nakanishi K, Tajima F, Nakamura A, Yagura S, Ookawara T, Yamashita H, Suzuki K, Taniguchi N, Ohno H (1995) Effects of hypobaric hypoxia on antioxidant enzymes in rats. J Physiol 489(Pt 3):869–876

van Patot MCT, Leadbetter G, Keyes LE, Bendrick-Peart J, Beckey VE, Christians U, Hackett P (2005) Greater free plasma VEGF and lower soluble VEGF receptor-1 in acute mountain sickness. J Appl Physiol (1985) 98(5):1626–1629

Peluso G, De Santis M, Inzitari R, Fanali C, Cabras T, Messana I, Castagnola M, Ferraccioli GF (2007) Proteomic study of salivary peptides and proteins in patients with Sjogren's syndrome before and after pilocarpine treatment. Arthritis Rheum 56:2216–2222

Peres R, Camargo G, Mofatto L et al (2010) Association of polymorphisms in the carbonic anhydrase 6 gene with salivary buffer capacity, dental plaque pH, and caries index in children aged 7–9 years. Pharmacogenomics J 10:114–119. https://doi.org/10.1038/tpj.2009.37

Pfaffe T, Cooper-White J, Beyerlein P, Kostner K, Punyadeera C (2011) Diagnostic potential of saliva: current state and future applications. Clin Chem 57:675–687

Pilardeau P, Richalet JP, Bouissou P, Vaysse J, Larmignat P, Boom A (1990) Saliva flow and composition in humans exposed to acute altitude hypoxia. Eur J Appl Physiol Occup Physiol 59: 450–453

Radak Z, Lee K, Choi W, Sunoo S, Kizaki T, Oh-ishi S, Suzuki K, Taniguchi N, Ohno H, Asano K (1994) Oxidative stress induced by intermittent exposure at a simulated altitude of 4000 m decreases mitochondrial superoxide dismutase content in soleus muscle of rats. Eur J Appl Physiol Occup Physiol 69:392–395

Radak Z, Asano K, Lee KC, Ohno H, Nakamura A, Nakamoto H, Goto S (1997) High altitude training increases reactive carbonyl derivatives but not lipid peroxidation in skeletal muscle of rats. Free Radic Biol Med 22:1109–1114

Rehak NN, Cecco SA, Csako G (2000) Biochemical composition and electrolyte balance of "unstimulated" whole human saliva. Clin Chem Lab Med 38:335–343

Reuter S, Gupta SC, Chaturvedi MM, Aggarwal BB (2010) Oxidative stress, inflammation, and cancer: how are they linked? Free Radic Biol Med 49:1603–1616

Ribon A, Pialoux V, Saugy JJ, Rupp T, Faiss R, Debevec T, Millet GP (2016) Exposure to hypobaric hypoxia results in higher oxidative stress compared to normobaric hypoxia. Respir Physiol Neurobiol 223:23–27

Rudney JD, Staikov RK, Johnson JD (2009) Potential biomarkers of human salivary function: a modified proteomic approach. Arch Oral Biol 54:91–100

San T, Polat S, Cingi C, Eskiizmir G, Oghan F, Cakir B (2013) Effects of high altitude on sleep and respiratory system and theirs adaptations. ScientificWorldJournal 2013:241569

Sartori C, Vollenweider L, Löffler B-M, Delabays A, Nicod P, Bartsch P, Scherrer U (1999) Exaggerated endothelin release in high-altitude pulmonary edema. Circulation 99:2665–2668

Schafer CA, Schafer JJ, Yakob M, Lima P, Camargo P, Wong DT (2014) Saliva diagnostics: utilizing oral fluids to determine health status. Monogr Oral Sci 24:88–98

Sedoris KC, Thomas SD, Miller DM (2010) Hypoxia induces differential translation of enolase/MBP-1. BMC Cancer 10:157

Semenza GL, Jiang BH, Leung SW, Passantino R, Concordet JP, Maire P, Giallongo A (1996) Hypoxia response elements in the aldolase a, enolase 1, and lactate dehydrogenase a gene promoters contain essential binding sites for hypoxia-inducible factor 1. J Biol Chem 271: 32529–32537

Sharp FR, Bernaudin M (2004) HIF1 and oxygen sensing in the brain. Nat Rev Neurosci 5:437–448

Siervo M, Riley HL, Fernandez BO, Leckstrom CA, Martin DS, Mitchell K, Levett DZH, Montgomery HE, Mythen MG, Grocott MPW, Feelisch M, for the Caudwell Xtreme Everest Research, G (2014) Effects of prolonged exposure to hypobaric hypoxia on oxidative stress, inflammation and Gluco-insular regulation: the not-so-sweet Price for good regulation. PLoS One 9:e94915

Sikri G, Bhattachar S (2017) Comment on "Soluble urokinase-type plasminogen activator receptor plasma concentration may predict susceptibility to high altitude pulmonary edema". Mediators Inflamm 2017:8546027

Singh M, Thomas P, Shukla D, Tulsawani R, Saxena S, Bansal A (2013) Effect of subchronic hypobaric hypoxia on oxidative stress in rat heart. Appl Biochem Biotechnol 169:2405–2419

Strapazzon G, Malacrida S, Vezzoli A, Dal Cappello T, Falla M, Lochner P, Moretti S, Procter E, Brugger H, Mrakic-Sposta S (2016) Oxidative stress response to acute hypobaric hypoxia and its association with indirect measurement of increased intracranial pressure: a field study. Sci Rep 6:32426

Streckfus CF, Bigler L (2016) A catalogue of altered salivary proteins secondary to invasive ductal carcinoma: a novel in vivo paradigm to assess breast cancer progression. Sci Rep 6:30800

Streckfus CF, Mayorga-Wark O, Arreola D, Edwards C, Bigler L, Dubinsky WP (2008) Breast cancer related proteins are present in saliva and are modulated secondary to ductal carcinoma in situ of the breast. Cancer Invest 26:159–167

Strubelt O, Younes M, Li Y (1994) Protection by albumin against ischaemia- and hypoxia-induced hepatic injury. Pharmacol Toxicol 75:280–284

Taylor AT (2011) High-altitude illnesses: physiology, risk factors, prevention, and treatment. Rambam Maimonides Med J 2:e0022

Van Molle W, Libert C, Fiers W, Brouckaert P (1997) Alpha 1-acid glycoprotein and alpha 1-antitrypsin inhibit TNF-induced but not anti-Fas-induced apoptosis of hepatocytes in mice. J Immunol 159:3555–3564

Van Nieuw Amerongen A, Bolscher JG, Veerman EC (2004) Salivary proteins: protective and diagnostic value in cariology? Caries Res 38:247–253

de Vries B, Walter SJ, Wolfs TG, Hochepied T, Rabina J, Heeringa P, Parkkinen J, Libert C, Buurman WA (2004) Exogenous alpha-1-acid glycoprotein protects against renal ischemia-reperfusion injury by inhibition of inflammation and apoptosis. Transplantation 78:1116–1124

Walsh NP, Laing SJ, Oliver SJ, Montague JC, Walters R, Bilzon JL (2004) Saliva parameters as potential indices of hydration status during acute dehydration. Med Sci Sports Exerc 36:1535–1542

Wilson MH, Newman S, Imray CH (2009) The cerebral effects of ascent to high altitudes. Lancet Neurol 8:175–191

Woods DR, Davison A, Stacey M, Smith C, Hooper T, Neely D, Turner S, Peaston R, Mellor A (2012) The cortisol response to hypobaric hypoxia at rest and post-exercise. Horm Metab Res 44:302–305

Xiao H, Zhang L, Zhou H, Lee JM, Garon EB, Wong DT (2012) Proteomic analysis of human saliva from lung cancer patients using two-dimensional difference gel electrophoresis and mass spectrometry. Mol Cell Proteomics 11(M111):012112

Xu RH, Pelicano H, Zhou Y, Carew JS, Feng L, Bhalla KN, Keating MJ, Huang P (2005) Inhibition of glycolysis in cancer cells: a novel strategy to overcome drug resistance associated with mitochondrial respiratory defect and hypoxia. Cancer Res 65:613–621

Yaari A, Tovbin D, Zlotnick M, Mostoslavsky M, Shemer-Avni Y, Hanuka N, Burbea Z, Katzir Z, Storch S, Margalith M (2006) Detection of HCV salivary antibodies by a simple and rapid test. J Virol Methods 133:1–5

Yamaguchi M, Takada R, Kambe S, Hatakeyama T, Naitoh K, Yamazaki K, Kobayashi M (2005) Evaluation of time-course changes of gingival crevicular fluid glucose levels in diabetics. Biomed Microdevices 7:53–58

Yoshikawa T, Furukawa Y, Wakamatsu Y, Takemura S, Tanaka H, Kondo M (1982) Experimental hypoxia and lipid peroxide in rats. Biochem Med 27:207–213

Yoshizawa JM, Schafer CA, Schafer JJ, Farrell JJ, Paster BJ, Wong DT (2013) Salivary biomarkers: toward future clinical and diagnostic utilities. Clin Microbiol Rev 26:781–791

Zalewska A, Zwierz K, Zolkowski K, Gindzienski A (2000) Structure and biosynthesis of human salivary mucins. Acta Biochim Pol 47:1067–1079

Zelles T, Purushotham KR, Macauley SP, Oxford GE, Humphreys-Beher MG (1995) Saliva and growth factors: the fountain of youth resides in us all. J Dent Res 74:1826–1832

Zhang L, Farrell JJ, Zhou H, Elashoff D, Akin D, Park NH, Chia D, Wong DT (2010) Salivary transcriptomic biomarkers for detection of resectable pancreatic cancer. Gastroenterology 138(949–57):e1–e7

Zhang Y, Duan R, Cui W, Pan Z, Liu W, Long C, Wang Y, Wang H (2013) Proteomic identification of human serum biomarkers associated with high altitude pulmonary edema. Zhongguo Ying Yong Sheng Li Xue Za Zhi 29:501–507

Zhou T, He Q, Tong Y, Zhan R, Xu F, Fan D, Guo X, Han H, Qin S, Chui D (2014) Phospholipid transfer protein (PLTP) deficiency impaired blood-brain barrier integrity by increasing cerebrovascular oxidative stress. Biochem Biophys Res Commun 445:352–356

Role of Genomics, Proteomics, and Antioxidant Interventions in Preventing High Altitude Sickness

Samakshi Verma and Arindam Kuila

Abstract

Some individuals become sick on spending some time at low atmospheric pressure which is above 1500 m (5000 feet), i.e. flying in a plane or climbing a mountain at high altitude. One of the three forms of high altitude sickness, i.e. AMS (acute mountain sickness having life frightening complications), HAPE (high altitude pulmonary edema, a non-cardiogenic form of pulmonary edema occurs due to excessive hypoxic pulmonary vasoconstriction which can be fatal if not recognized and treated promptly), and HACE (high altitude cerebral edema, a potentially fatal sickness characterized by ataxia, decreased consciousness, and characteristic changes on magnetic resonance imaging) may develop in an individual when they spend 1–5 days to altitudes $\geq$2500 m. Nowadays, there is an increasing population of visitors to ski resorts and mountains because of which milder forms of illness may occur at more moderate altitude. Problems experienced by a passenger when they fly in newer planes at 2440 m (8000 feet) which is higher equivalent cabin altitude than the earlier designs, can also be understood due to these research. In this chapter, the role of genomics, proteomics, and antibiotics interventions towards high altitude sickness is briefly discussed.

Keywords

High altitude · Acute mountain sickness · High altitude pulmonary edema · High altitude cerebral edema

S. Verma · A. Kuila (✉)
Department of Bioscience & Biotechnology, Banasthali Vidyapith, Banasthali, Rajasthan, India

© The Author(s), under exclusive license to Springer Nature Singapore Pte Ltd. 2022
N. K. Sharma, A. Arya (eds.), *High Altitude Sickness – Solutions from Genomics, Proteomics and Antioxidant Interventions*,
https://doi.org/10.1007/978-981-19-1008-1_11

11.1   Introduction

The condition hypobaric hypoxia which in some cases accelerates the chain of some physiological responses that will help the individual to adapt and tolerate the low availability of oxygen at high altitude but in other cases maladaptive responses takes place which in return causes AMS, HAPE or HACE (Luks et al. 2017). Among millions of people who visit high altitude every year, many of them suffer from AMS (acute mountain sickness), HACE (high altitude cerebral edema), and HAPE (high altitude pulmonary edema) which are all life threatening sickness with varying symptoms (Moore 1987).

The ambient pressure is higher at lower latitudes, and lower in summer as compared to winter. AMS (acute mountain sickness) increases with the low pressure weather because temporary weather patterns also affect pressure (Zafren and Honigman 1997). The heart rate is much higher with the faster climbing as compared to planes take off (McFarland 1953). Heart rates are related with higher altitude on mountains, i.e. the heart rate will increase by 52% at 9000 feet (2700 m) and 72% at 12,000 feet (3700 m) (Saito et al. 2000). The rate of breathing is increased continuously on climbing and it remains upraised (Zafren and Honigman 1997). There is also an increase in blood flow within the eye and the blood vessels present in the retina of eye also swell (Botella de Maglia and Martinez-Costa 1998; Frayser et al. 1970).

There are some changes that are not considered as pathological, i.e. disturbed breathing during sleep, increased urination even at night, shortness of breath during excretion, quick breathing, weird dreams are all normal at altitude but changed patterns of sleep can overtire the climbers (Dietz 2001). The person feels suffocated or awakes suddenly due to restlessness within the period when breathing ceases (Curtis 1995).

Oxygen shortage imputes decline in night vision which can further be restored within some minutes by inbreathing pure oxygen (Heath and Williams 1995). When people reach 8200–10,000 feet (2500–3000 m), they got fainted and they recover quickly and stand up immediately after eating without any further complications (Bezruchka 1994). On flights, fainting is most common medical incidents which contain 15–35% medical problems on boarding (Donaldson and Pearnt 1996; Harding and Mills 1993).

At 6500 feet (2000 m) altitude, the penetration of ultra-violet irradiation is 50% higher than at sea level which will ultimately increase the problems of eye damage and sunburns (Heath and Williams 1995). At 10,000 feet (3000 m) altitude faster climbing shows a remarkable effect in difficult mental tasks (McFarland 1953). At 5000 feet (1500 m), the reaction time was noteworthy high (22%) on investigation (Ernsting 1978). At 11,500 feet (3500 m), latency becomes high in the sounded invoke potential (Mukhopadhyay et al. 2000).

Above 2500 m altitude individuals get affected from AMS (acute mountain sickness). Figure 11.1 shows various symptoms of acute mountain sickness such as excessive flatulating, headache, loss of appetite, nose bleeds, palpitations, peripheral edema, persistent rapid pulse, pins and needles, shortness of breath, etc. (Roach

Fig. 11.1 Symptoms of acute
mountain sickness

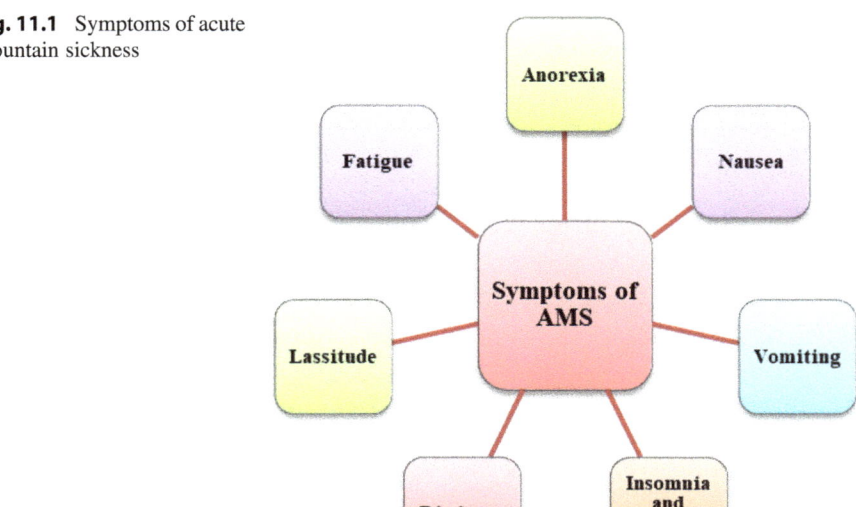

et al. 1996). The pathophysiology of AMS is uncertain (Kallenberg et al. 2007). The Lake Louise AMS symptom score and Environmental Symptoms Questionnaire help in measuring the severity and incidence of AMS (Roach et al. 1993). Every individual has different susceptibility towards AMS, few people may develop HACE and HAPE it is left untreated and some of them may not be highly affected by the primary symptoms.

It is recommended that acetazolamide (125–250 mg), two times a day starting from 24 h before climbing till few hours after subsiding to prevent AMS; although rest, immediate dropping and supplemental oxygen are known for best cure of AMS (Queiroz and Rapoport 2007). 750 mg of acetazolamide is most effective in treating AMS but acetazolamide (500 mg) was rejected for the treatment of AMS as it is ineffective. According to some researchers' dexamethasone is considered better for treating AMS, 8–16 mg of dexamethasone is equally effective as 750 mg of acetazolamide for treating AMS above 4000 m altitude (Dumont et al. 2000).

The maximum growth of pathological incidents that starts during AMS is known as HACE (high altitude cerebral edema). During AMS, there is an accumulation of extravascular fluid in brain which further increases due to climbing causing various symptoms (as shown in Fig. 11.2) like ataxia, coma, convulsions, and death. The major precaution to cure from AMS is taking rest at that altitude or immediate descent of at least 1000 m other than the pharmacological treatments.

The accumulation of extravascular fluid in alveolar airspaces of lung is known as HAPE (high altitude pulmonary edema). There are various symptoms of HAPE (as shown in Fig. 11.3) which includes tachycardia, dyspnea, tachypnea, cough, and pink frothy sputum associated with heavy breathing are the major stamp of HAPE. 30 mg PO of nifedipine every 12 h during climbing is the most commonly utilized

Fig. 11.2 Symptoms of high
altitude cerebral edema

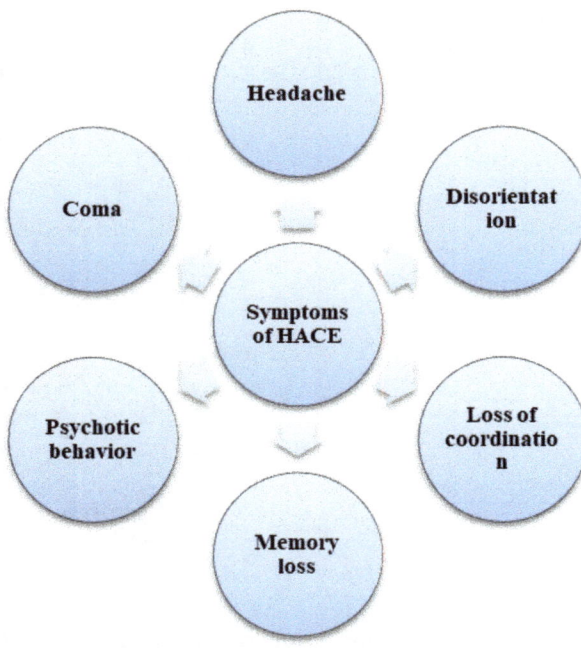

Fig. 11.3 Symptoms of high
altitude pulmonary edema

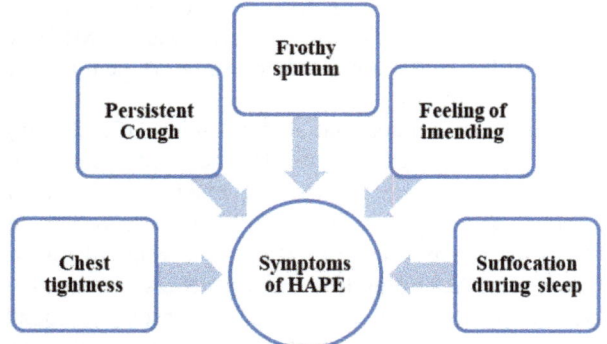

intervention for treating HAPE. Recent additions like tadalafil, with a longer half-life and sildenafil (both are phosphodiesterase inhibitors) are also utilized extensively. Nowadays, a new inhalable medicine is prescribed, i.e. Salmeterol (Pennardt 2013).

11.1.1 Role of Genetic Factors Towards High Altitude Sickness

There is no direct relation between susceptibility and gene polymorphisms till now and there is insubstantial data available about the role of genetic factors on high altitude sickness. Susceptibility of HAPE is linked with gene polymorphisms of endothelial nitric acid synthase only in Japan (Droma et al. 2002) but not in Europe (Weiss et al. 2003). There is no direct relation between ACE (angiotensin-converting enzyme) genes polymorphisms with susceptibility to HAPE but ACE gene polymorphisms may give out performance significance at high altitude (Dehnert et al. 2002a). According to the preliminary data, diseases like primary pulmonary hypertension and HAPE have different genetic bases but both disease susceptibilities share some common physiological relationship (Dehnert et al. 2002b).

Genome is not just a constant blueprint for organismal and cellular structure, function, and development on comparing human proteome and metabolome. There are several processes that play a crucial role in production to regulate alternate mRNA splicing, histone modifications, microRNAs, modulation of transcription factor activity, single nucleotide polymorphisms (SNPs), and tissue and response specific gene expression including promoter methylation (Kirby et al. 2007). In order to investigate the genome, technologies are improving with the greatest level of sophistication. High resolution analysis of chromosomal abnormalities, location analyses (to measure transcription factor binding activity), microRNA activity, SNPs, splice variants, and targeted and global analyses of DNA methylation patterns are all allowed and investigated by bead-based and microarray platforms, expression analyses. Due to hypoxia, there are various changes in gene expression (Storey 2006). Because of hypoxia, the gene expression of several microRNAs is upregulated (Kulshreshtha et al. 2007). Several functional classes show impact on expression of genes by microRNAs, i.e. apoptotic signaling and cell cycle regulation. In the hypoxic pathways, traditionally hypoxic genes were examined but with the use of these techniques several other new genes were also identified which was not implicated previously. As per this study, strength of genomic mechanisms is illustrated like the potential of analyzing various related and non-related molecules at the same time with various conditions. Key components can easily be identified and biochemical processes can be explained. All such limitations will be undoubtedly resolved with the growth of these powerful technologies.

11.1.2 Role of Proteomics Towards High Altitude Sickness

Two basic workflows are being generally focused by the study of proteomics profiling: (1) to resolve and relatively quantitate proteins with mass spectrometry known as 2DE (two-dimensional gel electrophoresis) or (2) metabolic and post-metabolic labeling and non-labeling methods known as quantitative mass spectrometry (Bowler et al. 2006). Either a light tag which is hydrogen or a heavy tag which is deuterium are the two samples which are being used in labeling of peptides in one mode of post-metabolic labeling. Mass spectrometry is used to measure the amount of relative peptide and to detect the mass difference between these two peptides while proteins are also being identified simultaneously. Cells are cultured in media which contains variants of amino acid within the typical metabolic labeling, e.g. 13C/15N-labeled arginine, which is incorporated into the entire cell proteome. Three states of arginine can be compared and their three different masses can be utilized such as Arg10 (13C6 15N4), Arg6 (13C6 14N4), and Arg0 (12C6 14N4). Several alternatives of these systems already existed which comprise the involvement of back end and front end fractionation/separation of proteins and peptides. Eventually, proteins are being identified with the help of mass analysis and upcoming database search. Significance of these workflows is that they have the strength to detect high amount of proteins without depending on gel-based methods which are discounted several protein classes, i.e. proteins found in membrane proteins and low abundance. Urine (low molecular weight and high salt protein) found in bio-fluids can make 2DE an unacceptable method. There are some consistent results which are observed between 2DE and labeling methods (Agarwal et al. 1995).

For clinical assays, there are several methodologies that have been used but there is no development of such method that can successfully analyze the whole proteome, i.e. urine proteome or human plasma and there is no validation of biomarkers for hypoxia till now. Focusing on the dynamic nature of human bio-fluid proteomes for more longitudinal studies, the available data on large-scale study of proteomics could be interpreted (Ommen 2005; Adachi et al. 2006). Only limitation related with proteomics is that there is a comprehensive shortage of sensitivity of the currently accepted analysis. Targeted analyses depending on antibodies which are sensitive but are difficult to develop into high yielding screens at the time of lack of antibodies and the selectivity of methods may sometimes not be appropriately examined (Anderson 2005). Analysis that depends on single biomarkers may not be as productive as those which contain panels of metabolites and proteins. Panels help in reducing the effect of normal variations between individuals but panels are also less vulnerable to noise than the single protein biomarkers.

There is a solution to overcome this problem is the use of mass spectrometry technique which is most used nowadays for analysis of small molecule known as multiple reaction monitoring (MRM). MRM consists of 13N and 15N stable isotope labeled artificial peptide as internal standards, quantitation of peptides, expected proteins, using mass spectrometry. The areas beneath peaks of endogenous and labeled peptides are analyzed, showing quantitative results and labeled and endogenous peptides are all analyzed by MRM. Antibody-based assays are not capable in differentiating between protein isoforms but MRM is more accurate method. Apo lipoprotein A1 with CV <4% is quantified by Barr et al. (1996); larger scale proteomics study has been done by (Gygi et al. 1999); and C-reactive protein in patients with rheumatoid arthritis was analyzed by Kuhn et al. (2004). A detailed review of MRM technique can be studied from Wright et al. (2005).

Immunoassays are highly specific and sensitive and antibody arrays (restricted by the quantity, quality, and availability of antibodies) which may be utilized for such high yielding screens (Hultschig et al. 2006). If it is unknown that either the proteins are capable biomarkers then it is restricted to develop antibody-based assays for large number of proteins in terms of money and time.

11.1.3   Role of Antioxidants Towards High Altitude Sickness

Phyto-extracts are most commonly used antioxidant against hypobaric hypoxia to enervate the oxidative stress caused by it at high altitude. Natural origins become favorite for the researchers because of their several qualities like biocompatibility without the need of much safety testing at high doses, easy availability, no need of medical supervision, and decreased requirement for synthesis. There are various phyto-extracts with significant effects on redox stress within the hypobaric hypoxia manifestation including, but are not restricted to, *Bacopa monniera* extracts (Hota et al. 2009), *Ginkgo biloba* extracts (Karcher et al. 1984; Oberpichler et al. 1988; Gertsch et al. 2004; Moraga et al. 2007; Hoyer et al. 1999), *quercetin* (Pandey et al. 2012), and *Withania somnifera* (Bhattacharya and Muruganandam 2003; Baitharu et al. 2013). Some other substances from natural or synthetic origin that are efficient against hypobaric hypoxia influenced oxidative stress are alphalipoic acid, alpha-ketoglutaric acid (Bailey and Davies 2001), acetyl-L-carnitine (Devi et al. 2007; Barhwal et al. 2009), ascorbic acid (Farias et al. 2010), Ceftriaxone (Hota et al. 2008), *Gabapentin* (Jafarian et al. 2008; Kumar and Goyal 2008), and 5-hydroxymethylfurfural (Gatterer et al. 2013; Zhang et al. 2015).

Problems with current pharmacological interventions like contra-indications in individuals with ongoing cardiac, hepatic, and renal conditions, cost, risks of over-dosing and medical supervision for use, and side effects are all can be controlled with the use of phyto-extracts. A novel stable aqueous suspension of micronized silymarin (SMN) is developed by all such efforts done. There are no side effects of SMN that are known and it has tolerance values reaching higher than 10 g/kg (oral; dogs) and 300 mg/kg (single IV bolus; dogs) (Arya et al. 2015). This dosage of SMN is reported safe for humans with 35 formulations (complex with higher cost) that are already marketed as a hepato-protective oral supplement having no problems related with cardiac and renal insufficiencies (Dixit et al. 2007). The rodent lung was protected from HH-induced oxidative stress with the usage of SMN and gets inflammated at 50 mg/kg/day with single oral doses repeated over 5 days (Paul et al. 2016). According to the previous study which shows several benefits and efficiency of SMN against HH-induced oxidative stress and inflammation, it is proven that SMN can be marketed in place of other current pharmacological interventions specifically in those individuals who are suffering from hepatic and renal problems.

11.2   Conclusion

Recently, development of powerful technologies takes place for inexpensive, large-scale producing, and widely utilized process for discovery. But according to the past research, the application and development of such process have been reported complicated for practical use which includes various reasons such as high variability seen in metabolomes of human populations, cost, and lack of comprehensive technical platforms which give amazing data sets. Modern analysis methods have several limitations like high cost related with cutting-edge techniques, lack of validation, low sample numbers, poor reproducibility. Additionally, experimental variability effects have not been globally identified. In order to apply and integrate such technologies to relevant disease states that there is a need of well-characterized samples from specifically stratified patient cohorts, which in exchange requires strong association between the basic scientist and clinician. If all these things are in place, it is more challenging to perform bioinformatics and statistical analyses because of examining multiple data sets simultaneously.

The only treatment of high altitude sickness is immediate descent from the altitude, rest at that altitude, and supplemental oxygen to prevent from any form of HAI (high altitude illness). Undoubtedly, high altitude illness (HAI) is a present known danger to all the visitors who move towards high altitude. Interventions utilized to treat high altitude sickness are efficient but they remain mounted with side effects and contra-indications.

References

Adachi J, Kumar C, Zhang Y, Olsen JV, Mann M (2006) The human urinary proteome contains more than 1500 proteins, including a large proportion of membrane proteins. Genome Biol 7(9): 1–16

Agarwal VR, Bulun SE, Simpson ER (1995) Quantitative detection of alternatively spliced transcripts of the aromatase cytochrome P450 (CYP19) gene in aromatase-expressing human cells by competitive RT-PCR. Mol Cell Probes 9(6):453–464

Anderson L (2005) Candidate-based proteomics in the search for biomarkers of cardiovascular disease. J Physiol 563(1):23–60

Arya A, Meena R, Sethy NK, Das M, Sharma M, Bhargava K (2015) NAP (davunetide) protects primary hippocampus culture by modulating expression profile of antioxidant genes during limiting oxygen conditions. Free Radic Res 49(4):440–452

Bailey DM, Davies B (2001) Acute mountain sickness; prophylactic benefits of antioxidant vitamin supplementation at high altitude. High Alt Med Biol 2(1):21–29

Baitharu I, Jain V, Deep SN, Hota KB, Hota SK, Prasad D, Ilavazhagan G (2013) Withania somnifera root extract ameliorates hypobaric hypoxia induced memory impairment in rats. J Ethnopharmacol 145(2):431–441

Barhwal K, Hota SK, Jain V, Prasad D, Singh SB, Ilavazhagan G (2009) Acetyl-l-carnitine (ALCAR) prevents hypobaric hypoxia-induced spatial memory impairment through extracellular related kinase-mediated nuclear factor erythroid 2-related factor 2 phosphorylation. Neuroscience 161(2):501–514

Barr JR, Maggio VL, Patterson DG Jr, Cooper GR, Henderson LO, Turner WE, Smith SJ, Hannon WH, Needham LL, Sampson EJ (1996) Isotope dilution—mass spectrometric quantification of specific proteins: model application with apolipoprotein AI. Clin Chem 42(10):1676–1682

Bezruchka S (1994) Altitude illness, prevention and treatment. Cordee, Leicester

Bhattacharya SK, Muruganandam AV (2003) Adaptogenic activity of Withania somnifera: an experimental study using a rat model of chronic stress. Pharmacol Biochem Behav 75(3): 547–555

Botella de Maglia J, Martinez-Costa R (1998) High altitude retinal hemorrhages in the expeditions to 8,000 meter peaks. A study of 10 cases. Med Clin 110(12):457–461

Bowler RP, Ellison MC, Reisdorph N (2006) Proteomics in pulmonary medicine. Chest 130(2): 567–574

Curtis R (1995) Outdoor action guide to high altitude: acclimatization and illnesses. Princeton, Outdoor Action Program

Dehnert C, Weymann J, Montgomery HE, Woods D, Maggiorini M, Scherrer U, Gibbs JSR, Bärtsch P (2002a) No association between high-altitude tolerance and the ACE I/D gene polymorphism. Med Sci Sports Exerc 34(12):1928–1933

Dehnert C, Miltenberger-Miltenyi G, Grunig E (2002b) Normal BMPR-2 gene in individuals susceptible to high altitude pulmonary edema. High Alt Med Biol 3:100

Devi SA, Vani R, Subramanyam MVV, Reddy SS, Jeevaratnam K (2007) Intermittent hypobaric hypoxia-induced oxidative stress in rat erythrocytes: protective effects of vitamin E, vitamin C, and carnitine. Cell Biochem Funct 25(2):221–231

Dietz TE (2001) All about altitude illness. Emergency and wilderness medicine. https://www.high-altitude-medicine.com/AMS.html. Accessed 21 February 2002

Dixit N, Baboota S, Kohli K, Ahmad S, Ali J (2007) Silymarin: a review of pharmacological aspects and bioavailability enhancement approaches. Indian J Pharm 39(4):172

Donaldson E, Pearnt J (1996) First aid in the air. Aust N Z J Surg 66(7):431–434

Droma Y, Hanaoka M, Ota M, Katsuyama Y, Koizumi T, Fujimoto K, Kobayashi T, Kubo K (2002) Positive association of the endothelial nitric oxide synthase gene polymorphisms with high-altitude pulmonary edema. Circulation 106(7):826–830

Dumont L, Mardirosoff C, Tramèr MR (2000) Efficacy and harm of pharmacological prevention of acute mountain sickness: quantitative systematic review. BMJ 321(7256):267–272

Ernsting J (1978) Prevention of hypoxia—acceptable compromises. Aviat Space Environ Med 49(3):495

Farias JG, Puebla M, Acevedo A, Tapia PJ, Gutierrez E, Zepeda A, Calaf G, Juantok C, Reyes JG (2010) Oxidative stress in rat testis and epididymis under intermittent hypobaric hypoxia: protective role of ascorbate supplementation. J Androl 31(3):314–321

Frayser R, Houston CS, Bryan AC, Rennie ID, Gray G (1970) Retinal hemorrhage at high altitude. N Engl J Med 282(21):1183–1184

Gatterer H et al (2013) Short-term supplementation with alpha-ketoglutaric acid and 5-hydroxymethylfurfural does not prevent the hypoxia induced decrease of exercise performance despite attenuation of oxidative stress. Int J Sports Med 34(01):1–7

Gertsch JH, Basnyat B, Johnson EW, Onopa J, Holck PS (2004) Randomised, double blind, placebo controlled comparison of ginkgo biloba and acetazolamide for prevention of acute mountain sickness among Himalayan trekkers: the prevention of high altitude illness trial (PHAIT). BMJ 328(7443):797

Gygi SP, Rist B, Gerber SA, Turecek F, Gelb MH, Aebersold R (1999) Quantitative analysis of complex protein mixtures using isotope-coded affinity tags. Nat Biotechnol 17(10):994–999

Harding RM, Mills FJ (eds) (1993) Aviation medicine. BMJ, London

Heath D, Williams DR (1995) High-altitude medicine and pathology. Oxford University Press, Oxford

Hota SK, Barhwal K, Ray K, Singh SB, Ilavazhagan G (2008) Ceftriaxone rescues hippocampal neurons from excitotoxicity and enhances memory retrieval in chronic hypobaric hypoxia. Neurobiol Learn Mem 89(4):522–532

Hota SK, Barhwal K, Baitharu I, Prasad D, Singh SB, Ilavazhagan G (2009) Bacopa monniera leaf extract ameliorates hypobaric hypoxia induced spatial memory impairment. Neurobiol Dis 34(1):23–39

Hoyer S, Lannert H, Nöldner M, Chatterjee SS (1999) Damaged neuronal energy metabolism and behavior are improved by Ginkgo biloba extract (EGb 761). J Neural Transm 106(11–12):1171–1188

Hultschig C, Kreutzberger J, Seitz H, Konthur Z, Büssow K, Lehrach H (2006) Recent advances of protein microarrays. Curr Opin Chem Biol 10(1):4–10

Jafarian S, Abolfazli R, Gorouhi F, Rezaie S, Lotfi J (2008) Gabapentin for prevention of hypobaric hypoxia-induced headache: randomized double-blind clinical trial. J Neurol Neurosurg Psychiatry 79(3):321–323

Kallenberg K, Bailey DM, Christ S, Mohr A, Roukens R, Menold E, Steiner T, Bärtsch P, Knauth M (2007) Magnetic resonance imaging evidence of cytotoxic cerebral edema in acute mountain sickness. J Cereb Blood Flow Metab 27(5):1064–1071

Karcher L, Zagermann P, Krieglstein J (1984) Effect of an extract of Ginkgo biloba on rat brain energy metabolism in hypoxia. Naunyn Schmiedebergs Arch Pharmacol 327(1):31–35

Kirby J, Heath PR, Shaw PJ, Hamdy FC (2007) Gene expression assays. Adv Clin Chem 44:247–292

Kuhn E, Wu J, Karl J, Liao H, Zolg W, Guild B (2004) Quantification of C-reactive protein in the serum of patients with rheumatoid arthritis using multiple reaction monitoring mass spectrometry and 13C-labeled peptide standards. Proteomics 4(4):1175–1186

Kulshreshtha R, Ferracin M, Wojcik SE, Garzon R, Alder H, Agosto-Perez FJ, Davuluri R, Liu CG, Croce CM, Negrini M, Calin GA, Ivan M (2007) A microRNA signature of hypoxia. Mol Cell Biol 27(5):1859–1867

Kumar A, Goyal R (2008) Gabapentin attenuates acute hypoxic stress–induced behavioral alterations and oxidative damage in mice: possible involvement of GABAergic mechanism. Indian J Exp Biol 46(3):159–163

Luks AM, Swenson ER, Bärtsch P (2017) Acute high-altitude sickness. Eur Respir Rev 26(143): 160096

McFarland RA (1953) Human factors in air transportation: occupational health and safety. McGraw-Hill, New York

Moore LG (1987) Altitude-aggravated illness: examples from pregnancy and prenatal life. Ann Emerg Med 16(9):965–973

Moraga FA, Flores A, Serra J, Esnaola C, Barriento C (2007) Ginkgo biloba decreases acute mountain sickness in people ascending to high altitude at Ollagüe (3696 m) in northern Chile. Wilderness Environ Med 18(4):251–257

Mukhopadhyay S, Thakur LALAN, Anand JP, Selvamurthy W (2000) Effect of sojourn at altitude of 3,500 m on auditory evoked potential in man. Indian J Physiol Pharmacol 44(2):211–214

Oberpichler H, Beck T, Abdel-Rahman MM, Bielenberg GW, Krieglstein J (1988) Effects of Ginkgo biloba constituents related to protection against brain damage caused by hypoxia. Pharmacol Res Commun 20(5):349–368

Ommen SR (2005) There is much more to the recipe than just outflow obstruction. J Am Coll Cardiol 46(8):1551–1552

Pandey AK, Patnaik R, Muresanu DF, Sharma A, Sharma HS (2012) Quercetin in hypoxia-induced oxidative stress: novel target for neuroprotection. Int Rev Neurobiol 102:107–146

Paul S, Arya A, Gangwar A, Bhargava K, Ahmad Y (2016) Size restricted silymarin suspension evokes integrated adaptive response against acute hypoxia exposure in rat lung. Free Radic Biol Med 96:139–151

Pennardt A (2013) High-altitude pulmonary edema: diagnosis, prevention, and treatment. Curr Sports Med Rep 12(2):115–119

Queiroz LP, Rapoport AM (2007) High-altitude headache. Curr Pain Headache Rep 11(4):293–296

Roach RC, Bärtsch P, Hackett PH, Oelz O (1993) The Lake Louise Acute Mountain Sickness Scoring System. In: Sutton JR, Houston CS, Coates G (eds) Hypoxia and molecular medicine. Queen City Press, Burlington, pp 272–274

Roach RC, Loeppky JA, Icenogle MV (1996) Acute mountain sickness: increased severity during simulated altitude compared with normobaric hypoxia. J Appl Physiol 81(5):1908–1910

Saito S, Aso C, Kanai M, Takazawa T, Shiga T, Shimada H (2000) Experimental use of a transportable hyperbaric chamber durable for 15 psi at 3700 meters above sea level. Wilderness Environ Med 11(1):21–24

Storey KB (2006) Genomic and proteomic approaches in comparative biochemistry and physiology. Physiol Biochem Zool 79(2):324–332

Weiss J, Haefeli WE, Gasse C, Hoffmann MM, Weyman J, Gibbs S, Mansmann U, Bärtsch P (2003) Lack of evidence for association of high altitude pulmonary edema and polymorphisms of the NO pathway. High Alt Med Biol 4(3):355–366

Wright ME, Han DK, Aebersold R (2005) Mass spectrometry-based expression profiling of clinical prostate cancer. Mol Cell Proteomics 4(4):545–554

Zafren K, Honigman B (1997) High-altitude medicine. Emerg Med Clin North Am 15(1):191–222

Zhang J-H, Di Y, Wu L-Y, He Y-L, Zhao T, Huang X, Ding X-F, Wu K-W, Fan M, Zhu L-L (2015) 5-HMF prevents against oxidative injury via APE/Ref-1. Free Radic Res 49(1):86–94

High Altitude Sickness and Antioxidant Interventions

12

Sarika Singh

Abstract

High altitude sickness (HAS) has been one of the important environmental challenges occurring as a result of the failure of the physiological acclimatization to acute hypobaric hypoxia. The prime bodily retort to hypoxia and adaptation linked to HAS comprises hyperventilation, increased systemic blood pressure together with tachycardia (increased heart rate) and increased hemoglobin concentration. These collective retorts enhance the supply of oxygen to the cells by means of alterations in the respiratory, cardiovascular, and hematologic system, thereby boosting the cellular oxygen uptake together with consumption mechanisms. Additionally, in majority of the cases, these bodily responses may not be sufficient, as a result the mount to high elevations and the associated hypoxia culminate in complex medical condition. Diverse strategies for preventing HAS can be employed categorized mainly into pharmacological and non-pharmacological and miscellaneous. Pharmacological interventions comprise drugs like acetazolamide or dexamethasone which have been found to work effectively but also linked with adverse physical secondary effects. Non-pharmacological and miscellaneous interventions are those which are not founded on the administration of drugs and can be categorized into two sets: pre-acclimatization and supplements. Pre-acclimatization and additional methods founded on pressure comprise the employment of hypobaric air breathing to sham higher elevations, positive end-expiratory pressure, and remote ischemic preconditioning. In addition, non-drug approaches, herbs, and natural supplements have been also tested and found to be promising in combating this challenge. Supplements may contain herbal extracts (like *Ginkgo biloba*,

S. Singh (✉)
School of Earth Sciences, Banasthali Vidyapith, Banasthali, Rajasthan, India

Rhodiola species, and Coca leaf products), mineral elements (like iron, magnesium), antacids, and hormonal agents (like medroxyprogesterone and erythropoietin). Hypoxia is well known for causing prooxidant/antioxidant disturbances in the cell, thereby resulting in oxidative stress. The key mechanisms through which hypoxia-induced reactive oxygen species (ROS) overgeneration takes place are increased catecholamine manufacture, mitochondrial redox potential diminution together with the stimulation of xanthine oxidase pathway. The scientific observations of a raised risk of ROS and free radical mediated oxidative insult at higher elevation have shown the way to the researchers to suggest nutritional antioxidant interventions such as ascorbic acid, beta-carotene, vitamin E, selenium, alpha-lipoic acid, etc. present in fruits and vegetables and herbal supplements rich in antioxidants like *Ginkgo biloba*, which may be advantageous in combating the problem of HAS. Several antioxidants and their concoction have been tested and have shown mixed results in relation to the effectiveness of antioxidants in averting HAS. Despite the bulk of scientific investigations exploring the beneficial effects as well as the effectiveness of various antioxidants against HAS in different experimental models, the present state of evidence is unable to substantiate or repudiate the use of any antioxidant or a concoction of antioxidant as a definite prophylactic agent. Additionally, an inadequate number of investigations support the effectiveness of antioxidant supplements at higher elevations in easing the difficulty of HAS. Furthermore, there is a dearth of standardized studies in this regard. The variations in the current investigations with regard to experimental design considerations like altitude reached, rate of ascension, degree of pre-acclimatization; antioxidant employed, purity and quality of the antioxidant, the timing, duration, pretreatment, and dosage administered; individual susceptibility, small sample size, etc. cloud the understanding with respect to their efficacy in bringing out the desirable effects. Another challenge in this regard is to identify the appropriate dosage, timing, or combination of antioxidants which will regulate the induced oxidative stress without any deleterious effects on the body and therefore help the body to ease out the symptoms of HAS. Therefore, given the undesirable side effects associated with pharmacological interventions, non-pharmacological antioxidant interventions which are relatively safer and with minimal side effects need more attention in order to combat this challenge of HAS.

Keywords

High altitude sickness · Antioxidants

12.1   Introduction

High altitude sickness (HAS) has been one of the important environmental challenges being faced by mankind since prehistoric times. It mainly occurs as a result of the failure of the physiological acclimatization to acute hypobaric hypoxia.

A huge number of individuals mount to higher elevations for various reasons ranging from occupational deployment, recreation, athletic competition, hobby to name a few. Apart from a considerable reduction in ambient temperature and humidity, the key environmental attributes defining the higher elevations are a significant plunge in barometric pressure resulting in a diminishment in the partial pressure of oxygen at every single point alongside the oxygen transport cascade commencing with the ambient environment to the powerhouse of the cell. Despite the constant oxygen percentage in the ambient air at about 21%, there is a drop in the atmospheric pressure with rise in height, as a result less amount of oxygen is available to the body consequently leading to hypobaric hypoxia (Zafren 2014).

This hypobaric hypoxia generates an array of physiological retorts that enables the climber to endure and acclimatize to the state of diminished oxygen pressure. In cases, where the body is not able to acclimatize due to defective bodily responses, the body succumbs to some or the other forms of HAS. Further, when the rate of ascension exceeds the rate of bodily adaptation it culminates into HAS (Zafren 2014). This quickly rescindable condition is characterized by a group of brain and lung syndromes occurring in individuals as a result of ascend to altitudes greater than approximately 2500 m (above 7000 feet). HAS is considered of two types: acute mountain sickness (AMS) and chronic mountain sickness (CMS), also referred to as Monge's disease (Monge 1942).

The prime bodily retort to hypoxia and adaptation linked to HAS comprises hyperventilation, increased systemic blood pressure together with tachycardia (increased heart rate), and increased hemoglobin concentration (Palmer 2010). These collective retorts enhance the supply of oxygen to the cells by means of alterations in the respiratory, cardiovascular, and hematologic system, thereby boosting the cellular oxygen uptake together with consumption mechanisms (Palmer 2010). Additionally, in majority of the cases, these bodily responses may not be sufficient, as a result the mount to high elevations and the associated hypoxia culminate in complex medical condition (Palmer 2010; Luks et al. 2017), well known as HAS. It can arise at any time ranging from several hours of ascension to post 5 days of ascension with mild to deadly conditions depending upon several physical and biological factors.

12.2   Acute Mountain Sickness (AMS), High Altitude Cerebral Edema (HACE), and High Altitude Pulmonary Edema (HAPE)

AMS is a syndrome characterized primarily by neurological symptoms. Scientific studies have corroborated the elevation in rate of respiration coupled with raised hemoglobin levels with experience to low oxygen pressure at higher altitudes (Palmer 2010; Paralikar and Paralikar 2010; Luks et al. 2017). This reduced pressure and diminished oxygen concentration at higher altitudes generates a set of symptoms comprising of headache, dizziness, shortness of breath, anorexia, or nausea, vomiting, tiredness, and insomnia frequently termed as AMS (Palmer 2010). The most predominant symptom of AMS is headache which can be considered as a discrete clinical entity. If neglected, the illness may advance to life threatening high altitude cerebral edema (HACE) or pulmonary edema. In severe cases where the brain (HACE) is chiefly affected, symptoms like drowsiness, confusion, unconsciousness or ataxia, impaired motor control can transpire (Imray et al. 2010), whereas when the lungs (high altitude pulmonary edema or HAPE) are disturbed, symptoms like cough or breathlessness appear. If unaddressed, HACE may lead to death as an upshot of cerebral edema. HACE is generally headed by the signs and symptoms of AMS and is often considered as the dangerous form and culmination phase of AMS (Imray et al. 2010; Palmer 2010; Zafren 2014). Further, studies indicate a connection by way of intracranial hypertension amid these syndromes through which they may share a common pathophysiology (Davis and Hackett 2017; Luks et al. 2017). The Lake Louise Questionnaire or Environmental Symptoms Questionnaire can be employed to score the severity of AMS in mountaineers.

Pulmonary alveolar hypoxia is a common term for non-adapted individuals who rapidly travel to high altitudes. It is characterized by the diminished trans-alveolar fluid transport resulting in excessive fluid buildup in the alveoli overstating alveolar hypoxia. This further impedes the gas exchange process consequently instigating the pathological development of HAPE, which is deadliest kind of HAS (Stream and Grissom 2008; Imray et al. 2010). Moreover, HAPE is a condition developing due to hypoxia within 2–4 days of ascent to higher elevations wherein the lungs are distressed as a result of pulmonary edema leading to intricate chain of events. It is described by a set of symptoms such as cough, difficult breathing (dyspnea), and reduced exercise tolerance (Palmer 2010). It is uncommon post 1 week of adaptation to a specific elevation (Palmer 2010). Basically, it generates as a result of the failure of the biological mechanisms to prevent the entry of water into the airspace (Scherrer et al. 2010) resulting in hypoxic pulmonary hypertension which is the distinguishing feature of HAPE. According to Scherrer et al. (2010), the faulty pulmonary nitric oxide production; overstated manufacture of endothelin-1; inflated sympathetic stimulation; and deficient alveolar transepithelial sodium transport have been suggested as the possible mechanisms behind the arbitration of pulmonary hypertension.

According to the approximations, around 84% of individuals flying straight to 3860 m are disturbed by AMS (Murdoch 1996). In comparison to AMS, the danger of HACE and HAPE is considerably lesser, with approximations varying between 0.1 and 4.0% (Basnyat and Murdoch 2003). Therefore, HACE and HAPE are though uncommon diseases in comparison to AMS nonetheless they are more deadly, particularly as the elevation rises (Stream and Grissom 2008). Generally, higher the altitude and faster the climb, the higher the possibility of occurrence of HAS. Moreover, depending upon the sensitivity, different individuals respond differently. So, individual susceptibility is yet another risk factor playing an important role in the development of HAS. Apart from these, history of HAS, permanent habitation below 900 m, physical exertion in kids and adults, obesity, and coronary heart disease could be other plausible risk factors behind the development of HAS (Basnyat and Murdoch 2003; Dehnert and Bärtsch 2010).

As far as gender is concerned, both men and women appear to be similarly vulnerable, together, better physical health does not ensure protection against HAS (Palmer 2010). To add, individuals suffering from asthma are strictly recommended to keep their health under control prior to any exertion at higher elevations (CATMAT 2007). Numerous expiries at higher altitudes like on Mt. Everest and other high mountains can be accredited as an upshot of altitude sickness. Furthermore, in general, HAS is a benign disorder troubling the individuals from sea level in case of a ski vacation or climbing in the mountains. The figure of people traveling quickly to higher elevations for work or tourism is increasing day by day. Betterment in transportation facilities for high elevation regions has further contributed to the rising number. Consequently, HAS has emerged as a major environmental health threat and prophylactic interventions are needed to combat this situation.

12.3 Diverse Strategies for Preventing High Altitude Sickness (HAS)

Various interventions have been documented in the scientific literature to escape HAS, particularly AMS (CATMAT 2007; Luks et al. 2010; Seupaul et al. 2012; Zafren 2014). On the whole, the various interventions employed for averting HAS can be categorized into pharmacological and non-pharmacological and miscellaneous (Luks and Swenson 2008; Luks et al. 2010). The Committee to Advise on Tropical Medicine and Travel suggested an accord for HAS, elaborating the preclusion and treatment methods amid various other issues concerning this clinical state (CATMAT 2007).

12.3.1 Pharmacological Interventions

Researchers have also studied diverse pharmacologic interventions to check HAS (Zafren 2014). In this context, drugs like acetazolamide or dexamethasone have been found to work effectively for the deterrence of AMS. Acetazolamide being a diuretic hinders the enzyme carbonic anhydrase, thereby causing an upsurge in renal excretion of potassium and bicarbonate. The induction of metabolic acidosis is the main target behind this prophylactic treatment, thus continuing hyperventilation at higher elevations and augmenting systemic oxygenation that eases AMS symptoms. Acetazolamide is a diuretic that inhibits carbonic anhydrase, producing an increase in renal secretion of potassium and bicarbonate. The aim of prophylactic treatment is to induce metabolic acidosis, thereby maintaining hyperventilation at high altitude and enhancing systemic oxygenation, which decreases the symptoms of AMS. The therapeutic efficacy of steroids, explicitly dexamethasone, has been linked to its antioxidant properties and consequent capacity to uphold the vascular integrity of the blood–brain barrier (BBB) (Hackett et al. 1998), which seems to be principally vulnerable to oxidative insult (Halliwell 1992). Sumatriptan and gabapentin have also been shown to give positive results against AMS but need more research

(Seupaul et al. 2012). However, pharmaceutical prevention has been linked with adverse physical secondary effects, for instance, acetazolamide though effective but triggers complaints like paresthesias, dysgeusia, and diuresis. AMS being a common problem at higher elevations, easily accessible, as well as harmless prophylactic means are wanted.

12.3.2 Non-Pharmacological and Miscellaneous Interventions

Non-pharmacological and miscellaneous interventions are those which are not founded on the administration of drugs and can be categorized into two sets: pre-acclimatization and supplements. Pre-acclimatization and additional methods founded on pressure comprise the employment of hypobaric air breathing to sham higher elevations, positive end-expiratory pressure, and remote ischemic preconditioning (Burse and Forte 1988; Launay et al. 2004; Dehnert et al. 2014; Berger et al. 2017). The risk of HAS escalates with a non-adapted mountaineer mounting to higher elevations greater than 2500 m (Paralikar and Paralikar 2010). Moreover, with increased individual susceptibility, an individual may suffer from AMS even below 2500 m at intermediate elevations like 2100 m (Davis and Hackett 2017). Further, the ascent speed together with exertion are the chief modifiable intermediaries of AMS risk. Therefore, though tough and unviable, gradual ascent or regulated rate of ascension, with regard to meters mounted per day has been suggested as one of the best ways of escaping HAS (CATMAT 2007; Paralikar and Paralikar 2010; Luks et al. 2014; Zafren 2014). While scheduling the rate of ascension, the elevation at which mountaineers sleep is of prime significance than the elevation attained during arousal (Luks et al. 2014). In gradual ascent, the mountaineers, especially individuals lacking prior altitude exposure, evade swift mounting to elevations more than 3000 m. Usually, they halt and stay at 2500–3000 m for 2–3 nights before mounting higher, using an additional night for adaptation every 600–900 m in case of ongoing climb. The main components in acclimatization are directed at safeguarding the availability of oxygen to body tissues and organs by way of optimum oxygen tension of the arterial blood (Bärtsch and Saltin 2008). Climbing to higher altitudes during daytime and coming back to lower elevations for rest assist in acclimatization (CATMAT 2007). Hypobaric and hypoxia chambers which change the amount of fractional inspired oxygen (FIO_2) or positive end-expiratory pressure (PEEP) imitating the effects of acclimatization could be appealing and extensively acknowledged for high elevation mountaineers as they would save time, transportation and effort in staging locations (Burse and Forte 1988; Dehnert et al. 2014; Launay et al. 2004). Additional interventions founded on remote ischemia in order to safeguard the brain could restore injury from subsequent ischemic slurs, because of its influences on vasoactive and inflammatory pathways (Berger et al. 2017; Pérez-Pinzón et al. 1997). To add, prior experience also aids in better acclimatization to the prevailing conditions. The knowledge and cognizance of AMS among the climbers have also been found to be effective in diminishing the prevalence of AMS (Gaillard et al. 2004). According to Vardy et al. (2005) cognizance among the climbers about the true signs and symptoms along with the prevention methods for AMS deters its occurrence.

In addition to drugs, non-drug approaches, herbs, and natural supplements have been tested and found to be promising in combating this challenge. Supplements may contain herbal extracts (like *Ginkgo biloba*, *Rhodiola* species, and Coca leaf products), mineral elements (like iron, magnesium), antacids and hormonal agents (like medroxyprogesterone and erythropoietin) (Roncin et al. 1996; Dumont et al. 1999; Gertsch et al. 2002, 2004; Chow et al. 2005; Moraga et al. 2007; Leadbetter et al. 2009; van Patot et al. 2009; Panossian et al. 2010; Seupaul et al. 2012; Chiu et al. 2013; Talbot et al. 2011; Ke et al. 2013; Heo et al. 2014; Ren et al. 2015). Iron supplement has been also shown to act as a prophylactic agent in diminishing the frequency of AMS in case of swift ascension to higher elevations in healthy participants (Talbot et al. 2011). The shielding effects of iron supplements have been attributed to its influence on the pathological and physiological responses to hypoxia, specifically those produced due to iron deficit (Ren et al. 2015). Moreover, the ability of iron supplements in alleviating pulmonary hypertension which may be caused due to acute scarcity of iron has been suggested as the mechanism behind its protective effect (Burtscher et al. 2004). Iron is a vital ingredient for the formation of human blood, and individuals lacking iron suffer from a disease known as anemia. In order to meet the deficiency, intravenous iron solutions such as dextran iron, ferrous gluconate, and iron sucrose have been employed in medical settings. Among these, iron sucrose which is a multi-core of iron (III)-hydroxide in sucrose has been indicated to be comparatively safer with minimal allergy risk together with least occurrence of undesirable effects (Hörl 2007). Contradictory results have also been reached. A preliminary research conducted by Ren and his colleagues (2015) utilized iron sucrose as intravenous iron supplementation in order to determine its efficacy in preventing AMS. The study revealed that intravenous iron supplementation has no substantial shielding effect in averting AMS.

Hormonal supplements are supposed to elevate the hypoxic ventilatory responses by means of improving the oxygen saturation and diminishing the hematocrit level together with the induction of red blood cell formation (Heo et al. 2014; Milledge and Cotes 1985). Hypoxia-induced erythropoietin (EPO) production is closely linked with adaptation to higher elevations. However, the release commences after 1–2 days of altitude exposure and it takes weeks to cause a rise in hemoglobin. (Milledge and Cotes 1985). EPO which is a glycoprotein has been found to be helpful against AMS by triggering the manufacture of red blood cell in the body together with a reduction in plasma volume, thereby causing a rise in the arterial O_2 concentration (Heo et al. 2014).

12.4 Cellular Antioxidant System and its Alteration in Relation to HAS

Hypoxia is well known for causing prooxidant/antioxidant disturbances in the cell, thereby resulting in oxidative stress (Magalhães et al. 2004; Pialoux et al. 2009). The key mechanisms through which hypoxia-induced reactive oxygen species (ROS) overgeneration takes place are increased catecholamine manufacture, mitochondrial redox potential diminution together with the stimulation of xanthine oxidase pathway (Mazzeo et al. 1998; Kehrer and Lund 1994). The production of free radicals which are extremely energized molecules possessing either one or more than one unpaired electron in their atomic orbital (Halliwell 1994) may further be a part of the cause in the intricate pathophysiology of HAS (Bailey and Davies 2001). Free radicals are generated in the body as a result of the cellular metabolism utilizing molecular oxygen. Though physiologically vital in controlled amounts, but then again, when in surplus could trigger and promulgate membrane destabilization and cell injury. Studies have connected the free radicals in a range of pathologies, especially lung disease like respiratory distress syndrome and a range of central nervous system complaints produced as a result of neurodegeneration, ischemia, or trauma (Halliwell and Gutteridge 1999). Nonetheless, it remains to be elucidated that they are the reason or simply an outcome of disease pathology. The oxidative insult caused as a result of these free radicals is however deterred by the efficient antioxidant defense system equipped with several enzymatic and non-enzymatic antioxidants. However, this antioxidant defense system is overwhelmed during mounting to high elevation (Bailey and Davies 2001). Numerous investigations have documented increased markers of oxidative stress as a result of high altitude exposure (Vasankari et al. 1997; Chao et al. 1999; Pfeiffer et al. 1999; Bailey et al. 2001; Joanny et al. 2001). To support this, Bailey et al. (2001) have indicated marked rise in lipid peroxidation and intracellular myofiber proteins soon afterward rise to 5100 m height. These upsurges were predominantly noticeable in volunteers developing AMS. Additionally, studies have shown that individuals from higher elevations have lower GPX activity (Imai et al. 1995). Further, the activity and effectiveness of GPX depend on the fitness of the thiol system. Glutamylcysteinyl-glycine which is continually manufactured by the glutamyl cycle is one among the key thiol/antioxidant source of the cell. Exposure to higher elevations results in diminishment of reduced glutathione (GSH) concentrations and elevation in oxidized glutathione concentrations (Ilavazhagan et al. 2001; Joanny et al. 2001) indicating the diminution in antioxidant capacity. Apart from hypobaric hypoxia, additional environmental factors contributing to the cumulative load of oxidative stress at higher elevations include exposure to ultra violet radiations, cold, diet poor in antioxidants, etc. depicted in figure 1 (Bailey and Davies 2001). However, the injury caused due to hypoxia-induced oxidative stress at higher elevations can be lessened by an antioxidant intervention, thereby reducing the toll of HAS (Askew 2002).

12.5 Role of Antioxidant Interventions and Prophylactic Benefits in HAS

The scientific observations of a raised risk of ROS and free radical mediated oxidative insult at higher elevation have shown the way to the researchers to suggest nutritional antioxidant interventions such as ascorbic acid, beta-carotene, vitamin E, selenium, alpha-lipoic acid, etc. present in fruits and vegetables and herbal supplements rich in antioxidants like *Ginkgo biloba*, which may be advantageous in combating the problem of HAS (Bailey et al. 2001; Gertsch et al. 2002; Askew 2002; Moraga et al. 2007; Panossian et al. 2010; Talbot et al. 2011; Seupaul et al. 2012; Chiu et al. 2013; Ke et al. 2013; Heo et al. 2014; Ren et al. 2015). Several antioxidants and their concoction have been tested against HAS. Studies have obtained mixed results in relation to the effectiveness of antioxidants in averting HAS, some observed beneficial effects (Roncin et al. 1996; Bailey and Davies 2001; Ilavazhagan et al. 2001; Schmidt et al. 2002; Gertsch et al. 2002; Moraga et al. 2007; Lee et al. 2013a, b), whereas others did not report any benefit (Pfeiffer et al. 1999; Gertsch et al. 2004; Chow et al. 2005; Baillie et al. 2009; Ke et al. 2013; Chiu et al. 2013).

In this connection, herbal supplements, like *Ginkgo biloba*, possess powerful antioxidant effect and cause arterial vasodilation, indicative of a link with nitric oxide (NO) and potential in hemodynamic disorders reducing free radicals generation as a consequence of hypoxia (Kleijnen and Knipschild 1992). *Ginkgo biloba* is well recognized as a herbal supplement owing flavonoid-mediated antioxidant capacity that has been found to safeguard important processes and structures of the cell, for instance, Na/K-ATPase function, ATP generation by mitochondria, and lipid membranes from the onslaught of oxidative stress (Du et al. 1999; Pierre et al. 1999). Therefore, it has also been tested as a novel prophylactic instrument for the deterrence of AMS. Numerous scientific investigations have suggested the prophylactic use of *Ginkgo biloba* against AMS (Roncin et al. 1996; Gertsch et al. 2002; Moraga et al. 2007). Animal studies conducted on rat models have also revealed its efficacy in averting HAPE (Berg 2004). Studies have indicated that NO may be involved in the pathophysiology of AMS by facilitating hypoxia-incited cerebral vasodilation (Roach and Hackett 2001; Van Mil et al. 2002; Moraga et al. 2007) and *Ginkgo biloba* is suggested to act as an NO forager, thereby diminishing NO concentrations in the cell (Marcocci et al. 1994). Additionally, a research carried out by the group of Moraga et al. (2007) indicated the efficacy of *Ginkgo biloba* in moderating the occurrence of AMS and stimulates a rise in oxygen saturation when administered 24 h prior to the start of ascend as well as continually during ascend to higher altitudes in subjects with no prior experience of mountaineering. The diminution in NO generation as a result of the inhibition of the enzyme NO synthase facilitated by *Ginkgo biloba* could be the reason behind the decline in AMS symptoms, thus lessening cerebral perfusion and penetrability of the blood–brain barrier (Schilling and Wahl 1999). Moreover, *Ginkgo biloba* may also check the activity of the enzyme phosphodiesterase, consequently increasing the relaxation of parietal smooth muscles and causing vasodilation of parietal vessels which in line enhances tissue perfusion and reduces local hypoxia (Marcocci et al. 1994). Additional probable modes of actions of *Ginkgo biloba* may comprise rise in endogenous antioxidants, diminution in free radical generation together with diminished lung leak during hypoxia (Louajri et al. 2001; Naik et al. 2006).

In contrast, many studies have not been able to show the prophylactic effect of ginkgo against AMS symptoms (Gertsch et al. 2004; Chow et al. 2005; Ke et al. 2013). In a randomized, double-blind, placebo-controlled study conducted by Gertsch et al. (2004) on 614 healthy volunteers receiving ginkgo, acetazolamide, acetazolamide and ginkgo in concoction, or placebo, did not report any significant effect of ginkgo in decreasing the count or austerity of AMS in comparison to placebo. Further, the concoction of acetazolamide and ginkgo resulted in a marginally significant diminution in the effectiveness of acetazolamide. The occurrence of AMS as documented by the researchers was 34% in control placebo group, 12% in acetazolamide, 35% for ginkgo, and 14% in ginkgo and acetazolamide concoction. Together, the fraction of patients with heightened austerity of AMS was found to be 18% for placebo, 3% for acetazolamide, 18% for ginkgo, and 7% for ginkgo and acetazolamide concoction. The authors anticipated the reason behind the failure of ginkgo as a prophylactic agent compared to preceding findings with positive effects to be its administration at the time of enrollment at a high baseline elevation contrasted with giving the medication at sea level prior to mounting. Additional explanations behind these differences could be the differences in experimental protocol and design; timing, duration and dosage administered; variation in altitude at which *Ginkgo biloba* is started; time of high altitude exposure along with quality and purity of *Ginkgo biloba*. In this line, Leadbetter et al. (2009) showed another cause for disparity to be the differences in the composition of the *Ginkgo biloba* extract. This group of researchers compared *Ginkgo biloba* extract from two varied sources and realized the difference in composition and their capacity to lessen the frequency and austerity of AMS. Zafren (2014) suggested the reason behind the differing outcomes reached in different investigations to the variance in constitution of Ginkgo since it lacks standard formulation, together with, the difference in its sources.

These conflicting evidences concerning the efficacy of *Ginkgo biloba* cloud our understanding. Recent reviews indicate the insufficiency of available data with regard to the effectiveness of *Ginkgo biloba* in safeguarding against AMS and thus more larger randomized controlled investigations are needed to fill the gaps (Tsai et al. 2018). With regard to the safety of *Ginkgo biloba*, the current scientific literature indicates that it is harmless, however, it is still not clear whether it bears the capacity to prevent AMS or not. Low dose of acetazolamide is presently the best prophylaxis against AMS. Given the composite nature and inconsistency of *Ginkgo biloba* extract, it is doubtful to be reliably active in averting AMS and hence cannot be suggested as a dependable prophylaxis for AMS (van Patot et al. 2009; Zafren 2014).

Another well-accepted phytoadaptogen employed against HAS is *Rhodiola* species. Genus *Rhodiola* belonging to the family Crassulaceae has been a treasured medicinal plant for combating HAS in Asian and European countries for thousands of years (Panossian et al. 2010; Hung et al. 2011; Lee et al. 2013a). Globally over 90 species are available and they are well recognized for their great antioxidant activities. Different species possess different amounts of the bioactive elements and therefore differ in their therapeutic usage in the native areas. Among all the recognized *Rhodiola* species, *Rhodiola crenulate* (*R. crenulata*) has been utilized as an antidote for AMS in Tibet since prehistoric times (Lee et al. 2013a, b). This species of *Rhodiola* was found to display great antioxidant activity and alleviated hypobaric hypoxia-induced pulmonary edema in rodents (Lee et al. 2013a). According to latest findings, Na/K-ATPase plays a significant part in the clearance of alveolar fluid. Studies have revealed that the inhibition as well as the knockdown of Na/K-ATPase expression appreciably diminished the clearance of alveolar fluid in rodent models (Icard and Saumon 1999; Looney et al. 2005). Salidroside and tyrosol have been identified as the two bioactive constituents of *R. crenulata* possessing antioxidant, anti-fatigue, anti-inflammatory, and anti-depression properties. The probable mode of action behind the ability of *R. crenulata* in mitigating pulmonary edema in rodent models could be the inhibition of hypoxia-induced Na/K-ATPase endocytosis and preservation of the integrity of alveolar-capillary barrier and pulmonary sodium transport (Lee et al. 2013a). Salidroside and tyrosol are considered as the key bioactive ingredients responsible for the effectiveness of *Rhodiola* species. In this regard, Lee et al. (2013b) through their study confirmed the antioxidant capacity of *Rhodiola crenulata* extract and its bioactive compounds tyrosol and salidroside in the control of Na, K-ATPase endocytosis as well as ROS generation in reaction to hypoxia through similar mechanisms, thereby indicating the important part played by these compounds in the shielding effects of *Rhodiola crenulata* in averting oxidative stress-related illnesses. Other researches have also exhibited beneficial effects of *Rhodiola* in preventing HAS in animal as well as human models (Lee et al. 2013b). Contrary scientific evidences have also been reached which clouds the understanding regarding the ability and efficacy of *Radiola* species in combating HAS. In this connection, Chiu et al. (2013) conducted a randomized, double-blind, placebo-controlled, crossover trial in healthy adult subjects randomized to two treatment series with 800 mg *R. crenulata* extract purified from the rhizome or placebo every day for 1 week prior to mounting and 2 days in the course of climbing, before crossing over to the alternative treatment afterwards a 3 month wash-out period. The researchers did not report any significant change in terms of the occurrence of AMS (as described by a Lake Louise score ≥ 3, counting headache together with no less than one of the indications of nausea or vomiting, fatigue, dizziness, or difficulty sleeping) between the two treatment series, i.e. individuals taking *Rhodiola*-placebo and those taking placebo-*Rhodiola* (all 60.8%; adjusted odds ratio (AOR) = 1.02, 95% confidence interval (CI) = 0.69–1.52). The prevalence of severe AMS in *Rhodiola* extract vs. placebo set was found to be 35.3% vs. 29.4% (AOR = 1.42, 95% CI = 0.90–2.25), thus indicating the ineffectiveness of *Rhodiola* extract in lessening the occurrence or severity of AMS.

Coca leaf, coca tea, and other coca-derivatives have been traditionally utilized by locals for preventing AMS (Salazar et al. 2012; Luks et al. 2014; Zafren 2014). According to anecdotal reports, coca-derivatives are nowadays also being employed by the mountaineers in Asian and African countries for the same (Luks et al. 2014). However, there is a shortage of scientific data proving its efficacy against AMS (Luks et al. 2010; Zafren 2014). In fact, mechanisms have been proposed through which it may raise the risk of AMS. The influence of catecholamine gush in the cardiovascular system may expound part of the pathophysiology (Salazar et al. 2012).

Antioxidants such as ascorbic acid possesses the ability to counteract aqueous superoxide, peroxyl, and alkoxyl radicals and can indirectly produce alpha-tocopherol (Bendich et al. 1986). The fat-soluble alpha-tocopherol is perhaps the most significant chain-breaking antioxidant having the capacity to forage peroxyl radicals. At the same time, alpha-lipoic acid which is both water and lipid soluble possesses the ability to reduce hydroxyl, peroxyl, ascorbyl, and chromanoxyl radicals. It also owns the power to increase the intracellular concentration of glutathione and restore ascorbate and alpha-tocopherol (Packer et al. 1997). Therefore, it is thought as a vital patron to the antioxidant defense mechanism. Bailey and Davies (2001) conducted a randomized double-blind placebo-controlled trial to examine the effectiveness of the concoction of water and lipid soluble antioxidant vitamins. The combination comprised of L-ascorbic acid, dl alpha-tocopherol acetate, together with alpha-lipoic acid, chiefly picked because of their capacity to check lipid peroxidation. The study reflected the physiological effectiveness of chronic antioxidant supplementation by reducing the prevalence as well as the severity of AMS, betterment in resting arterial oxygen saturation, and rise in caloric intake and further suggesting the influential part of oxygen free radicals in the pathophysiology. The authors attributed the decline in AMS prevalence to the diminution in the free radical facilitated cerebral edema as a result of the enhancement in vascular integrity of the BBB. This mode of action could also elucidate the findings of another placebo-controlled research which revealed relatively lower AMS scores after the treatment with *Ginkgo biloba* extract (Egb 761) (Roncin et al. 1996). An antioxidant concoction comprising vitamin E, beta-carotene, ascorbic acid, selenium, alpha-lipoic acid, N-acetyl 1-cysteine, catechin, lutein, and lycopene was also found to be effective in decreasing the oxidative injury caused due to high elevation (Schmidt et al. 2002). Findings from animal studies also stand in support of the potential of antioxidants in effectively reducing the cerebral/pulmonary edema induced experimentally in animal models (Armstead et al. 1992). Oral vitamin E supplementation in rats exposed to hypoxic exposure of 7576 m remarkably decreased the lipid peroxidation (Ilavazhagan et al. 2001). Baillie et al. (2009) on the contrary did not report any benefit of antioxidant intervention employing a concoction of L-ascorbic acid, alpha-tocopherol acetate and alpha-lipoic acid against diminishing the occurrence or austerity of AMS. Pfeiffer et al. (1999) also did not report any benefit from a mixture of antioxidants containing beta-carotene, vitamin E, vitamin C, selenium, and zinc in averting oxidative injury to the biomolecules.

Despite the bulk of scientific investigations exploring the beneficial effects as well as the effectiveness of various antioxidants against HAS in different experimental models, the present state of evidence is unable to substantiate or repudiate the use of any antioxidant or a concoction of antioxidant as a definite prophylactic agent. Additionally, an inadequate number of investigations support the effectiveness of antioxidant supplements at higher elevations in easing the difficulty of HAS. Furthermore, there is a dearth of standardized studies in this regard. The variations in the current investigations with regard to experimental design considerations like altitude reached, rate of ascension, degree of pre-acclimatization; antioxidant employed, purity and quality of the antioxidant, the timing, duration, pretreatment, and dosage administered; individual susceptibility, small sample size, etc. cloud the understanding with respect to their efficacy in bringing out the desirable effects. Another challenge in this regard is to identify the appropriate dosage, timing or combination of antioxidants which will regulate the induced oxidative stress without any deleterious effects on the body and therefore help the body to ease out the symptoms of HAS. Therefore, given the undesirable side effects associated with pharmacological interventions, non-pharmacological antioxidant interventions which are relatively safer and with minimal side effects need more attention in order to combat this challenge of HAS.

References

Armstead WM, Mirro R, Thelin OP, Shibata M, Zuckerman SL, Shanklin DR, Busija DW, Leffler CW (1992) Polyethylene glycol superoxide dismutase and catalase attenuate increased blood-brain barrier permeability after ischemia in piglets. Stroke 23:755–762

Askew EW (2002) Work at high altitude and oxidative stress: antioxidant nutrients. Toxicology 180(2):107–119. https://doi.org/10.1016/s0300-483x(02)00385-2

Bailey DM, Davies B (2001) Acute mountain sickness; prophylactic benefits of antioxidant vitamin supplementation at high altitude. High Alt Med Biol 2(1):21–29. https://doi.org/10.1089/152702901750067882

Bailey DM, Davies B, Young IS, Hullin DA, Seddon PS (2001) A potential role for free radical-mediated skeletal muscle soreness in the pathophysiology of acute mountain sickness. Aviat Space Environ Med 72(6):513–521

Baillie JK, Thompson AA, Irving JB, Bates MG, Sutherland AI, Macnee W, Maxwell SR, Webb DJ (2009) Oral antioxidant supplementation does not prevent acute mountain sickness: double blind, randomized placebo-controlled trial. QJM 102(5):341–348. https://doi.org/10.1093/qjmed/hcp026

Bärtsch P, Saltin B (2008) General introduction to altitude adaptation and mountain sickness. Scand J Med Sci Sports 18(Suppl 1):1–10. https://doi.org/10.1111/j.1600-0838.2008.00827.x

Basnyat B, Murdoch DR (2003) High-altitude illness. Lancet 361:1967–1974

Bendich A, Machlin LJ, Scandurra O, Burton GW, Wayner DDM (1986) The antioxidant role of vitamin C. Adv Free Radic Biol Med 2(2):419–444

Berg JT (2004) Ginkgo biloba extract prevents high altitude pulmonary edema in rats. High Alt Med Biol 5(4):429–434. https://doi.org/10.1089/ham.2004.5.429

Berger MM, Macholz F, Lehmann L, Dankl D, Hochreiter M, Bacher B et al (2017) Remote ischemic preconditioning does not prevent acute mountain sickness after rapid ascent to 3,450 m. J Appl Physiol 123(5):1228–1234

Burse RL, Forte VA Jr (1988) Acute mountain sickness at 4500 m is not altered by repeated eight-hour exposures to 3200-3550 m normobaric hypoxic equivalent. Aviat Space Environ Med 59(10):942–949

Burtscher M, Flatz M, Faulhaber M (2004) Prediction of susceptibility to acute mountain sickness by SaO_2 values during short-term exposure to hypoxia. High Alt Med Biol 5(3):335–340. https://doi.org/10.1089/ham.2004.5.335

Chao WH, Askew EW, Roberts DE, Wood SM, Perkins JB (1999) Oxidative stress in humans during work at moderate altitude. J Nutr 129(11):2009–2012. https://doi.org/10.1093/jn/129.11.2009

Chiu T-F, Chen LL-C, Su D-H, Lo H-Y, Chen C-H, Wang S-H, Chen W-L (2013) Rhodiola crenulata extract for prevention of acute mountain sickness: a randomized, double-blind, placebo-controlled, crossover trial. BMC Complement Altern Med 13:298. https://doi.org/10.1186/1472-6882-13-298

Chow T, Browne V, Heileson HL, Wallace D, Anholm J, Green SM (2005) Ginkgo biloba and acetazolamide prophylaxis for acute mountain sickness: a randomized, placebo-controlled trial. Arch Intern Med 165(3):296–301. https://doi.org/10.1001/archinte.165.3.296

Committee to Advise on Tropical Medicine and Travel (CATMAT) (2007) Statement on high-altitude illnesses. An Advisory Committee Statement (ACS). Can Commun Dis Rep 33 (ACS-5):1–20

Davis C, Hackett P (2017) Advances in the prevention and treatment of high altitude illness. Emerg Med Clin North Am 35(2):241–260. https://doi.org/10.1016/j.emc.2017.01.002

Dehnert C, Bärtsch P (2010) Can patients with coronary heart disease go to high altitude? High Alt Med Biol 11(3):183–188. https://doi.org/10.1089/ham.2010.1024

Dehnert C, Böhm A, Grigoriev I, Menold E, Bärtsch P (2014) Sleeping in moderate hypoxia at home for prevention of acute mountain sickness (AMS): a placebo-controlled, randomized double-blind study. Wilderness Environ Med 25(3):263–271. https://doi.org/10.1016/j.wem.2014.04.004

Du G, Willet K, Mouithys-Mickalad A, Sluse-Goffart CM, Droy-Lefaix MT, Sluse FE (1999) EGb 761 protects liver mitochondria against injury induced by in vitro anoxia/reoxygenation. Free Radic Biol Med 27(5-6):596–604. https://doi.org/10.1016/s0891-5849(99)00103-3

Dumont L, Mardirosoff C, Soto-Debeuf G, Tassonyi E (1999) Magnesium and acute mountain sickness. Aviat Space Environ Med 70(6):625

Gaillard S, Dellasanta P, Loutan L, Kayser B (2004) Awareness, prevalence, medication use, and risk factors of acute mountain sickness in tourists trekking around the Annapurnas in Nepal: a 12-year follow-up. High Alt Med Biol 5(4):410–419. https://doi.org/10.1089/ham.2004.5.410

Gertsch JH, Seto TB, Mor J, Onopa J (2002) Ginkgo biloba for the prevention of severe acute mountain sickness (AMS) starting one day before rapid ascent. High Alt Med Biol 3(1):29–37. https://doi.org/10.1089/152702902753639522

Gertsch JH, Basnyat B, Johnson EW, Onopa J, Holck PS (2004) and Prevention of High Altitude Illness Trial Research Group. Randomised, double blind, placebo controlled comparison of Ginkgo biloba and acetazolamide for prevention of acute mountain sickness among Himalayan trekkers: the prevention of high altitude illness trial (PHAIT). BMJ 328(7443):797. https://doi.org/10.1136/bmj.38043.501690.7C

Hackett PH, Yarnell PR, Hill R, Reynard K, Heit J, McCormick J (1998) High-altitude cerebral edema evaluated with magnetic resonance imaging: clinical correlation and pathophysiology. JAMA 280(22):1920–1925. https://doi.org/10.1001/jama.280.22.1920

Halliwell B (1992) Reactive oxygen species and the central nervous system. J Neurochem 59(5):1609–1623. https://doi.org/10.1111/j.1471-4159.1992.tb10990.x

Halliwell B (1994) Free radicals and antioxidants: a personal view. Nutr Rev 52(8 Pt 1):253–265. https://doi.org/10.1111/j.1753-4887.1994.tb01453.x

Halliwell B, Gutteridge JMC (1999) Free radicals in biology and medicine. Oxford University Press, Oxford

Heo K, Kang JK, Choi CM, Lee MS, Noh KW, Kim SB (2014) Prophylactic effect of erythropoietin injection to prevent acute mountain sickness: an open-label randomized controlled trial. J Korean Med Sci 29(3):416–422. https://doi.org/10.3346/jkms.2014.29.3.416

Hörl WH (2007) Iron therapy for renal anemia: how much needed, how much harmful? Pediatr Nephrol 22(4):480–489. https://doi.org/10.1007/s00467-006-0405-y

Hung SK, Perry R, Ernst E (2011) The effectiveness and efficacy of Rhodiola rosea L.: a systematic review of randomized clinical trials. Phytomedicine 18(4):235–244. https://doi.org/10.1016/j.phymed.2010.08.014

Icard P, Saumon G (1999) Alveolar sodium and liquid transport in mice. Am J Phys 277(6):L1232–L1238

Ilavazhagan G, Bansal A, Prasad D, Thomas P, Sharma SK, Kain AK, Kumar D, Selvamurthy W (2001) Effect of vitamin E supplementation on hypoxia-induced oxidative damage in male albino rats. Aviat Space Environ Med 72(10):899–903

Imai H, Kashiwazaki H, Suzuki T, Kabuto M, Himeno S, Watanabe C, Moji K, Kim SW, Rivera JO, Takemoto T (1995) Selenium levels and glutathione peroxidase activities in blood in an Andean high-altitude population. J Nutr Sci Vitaminol (Tokyo) 41(3):349–361. https://doi.org/10.3177/jnsv.41.349

Imray C, Wright A, Subudhi A, Roach R (2010) Acute mountain sickness: pathophysiology, prevention, and treatment. Prog Cardiovasc Dis 52(6):467–484. https://doi.org/10.1016/j.pcad.2010.02.003

Joanny P, Steinberg J, Robach P, Richalet JP, Gortan C, Gardette B, Jammes Y (2001) Operation Everest III (Comex '97): the effect of simulated sever hypobaric hypoxia on lipid peroxidation and antioxidant defence systems in human blood at rest and after maximal exercise. Resuscitation 49(3):307–314. https://doi.org/10.1016/s0300-9572(00)00373-7

Ke T, Wang J, Swenson ER, Zhang X, Hu Y, Chen Y, Liu M, Zhang W, Zhao F, Shen X, Yang Q, Chen J, Luo W (2013) Effect of acetazolamide and gingko biloba on the human pulmonary vascular response to an acute altitude ascent. High Alt Med Biol 14(2):162–167. https://doi.org/10.1089/ham.2012.1099

Kehrer JP, Lund LG (1994) Cellular reducing equivalents and oxidative stress. Free Radic Biol Med 17(1):65–75. https://doi.org/10.1016/0891-5849(94)90008-6

Kleijnen J, Knipschild P (1992) Ginkgo biloba. Lancet 340(8828):1136–1139. https://doi.org/10.1016/0140-6736(92)93158-j

Launay JC, Nespoulos O, Guinet-Lebreton A, Besnard Y, Savourey G (2004) Prevention of acute mountain sickness by low positive end-expiratory pressure in field conditions. Scand J Work Environ Health 30(4):322–326

Leadbetter G, Keyes LE, Maakestad KM, Olson S, Tissot van Patot MC, Hackett PH (2009) Ginkgo biloba does—and does not—prevent acute mountain sickness. Wilderness Environ Med 20(1):66–71. https://doi.org/10.1580/08-WEME-BR-247.1

Lee SY, Li MH, Shi LS, Chu H, Ho CW, Chang TC (2013a) Rhodiola crenulata extract alleviates hypoxic pulmonary edema in rats. Evid Based Complement Alternat Med 2013:718739, 9 p. https://doi.org/10.1155/2013/718739

Lee SY, Shi LS, Chu H, Li MH, Ho CW, Lai FY et al (2013b) Rhodiola crenulata and its bioactive components, salidroside and tyrosol, reverse the hypoxia-induced reduction of plasma-membrane-associated Na, K-ATPase expression via inhibition of ROS-AMPK-PKC ξ pathway. Evid Based Complement Alternat Med 2013:284150. https://doi.org/10.1155/2013/284150

Looney MR, Sartori C, Chakraborty S, James PF, Lingrel JB, Matthay MA (2005) Decreased expression of both the alpha1- and alpha2-subunits of the Na-K-ATPase reduces maximal alveolar epithelial fluid clearance. Am J Phys Lung Cell Mol Phys 289(1):L104–L110

Louajri A, Harraga S, Godot V, Toubin G, Kantelip JP, Magnin P (2001) The effect of Ginkgo biloba extract on free radical production in hypoxic rats. Biol Pharm Bull 24(6):710–712. https://doi.org/10.1248/bpb.24.710

Luks AM, Swenson ER (2008) Medication and dosage considerations in the prophylaxis and treatment of high-altitude illness. Chest 133(3):744–755. https://doi.org/10.1378/chest.07-1417

Luks AM, McIntosh SE, Grissom CK, Auerbach PS, Rodway GW, Schoene RB, Zafren K, Hackett PH (2010) Wilderness Medical Society Consensus Guidelines for the Prevention and Treatment of Acute Altitude Illness. Wilderness Environ Med 21:146–155

Luks AM, McIntosh SE, Grissom CK, Auerbach PS, Rodway GW, Schoene RB et al (2014) Wilderness Medical Society Practice Guidelines for the prevention and treatment of acute altitude illness: 2014 update. Wilderness Environ Med 25(4 Suppl):S4–S14

Luks AM, Swenson ER, Bärtsch P (2017) Acute high-altitude sickness. Eur Respir Rev 26:160096. https://doi.org/10.1183/16000617.0096-2016

Magalhães J, Ascensão A, Viscor G, Soares J, Oliveira J, Marques F, Duarte J (2004) Oxidative stress in humans during and after 4 hours of hypoxia at a simulated altitude of 5500 m. Aviat Space Environ Med 75(1):16–22

Marcocci L, Maguire JJ, Droy-Lefaix MT, Packer L (1994) The nitric oxide-scavenging properties of Ginkgo biloba extract EGb 761. Biochem Biophys Res Commun 201(2):748–755. https://doi.org/10.1006/bbrc.1994.1764

Mazzeo RS, Child A, Butterfield GE, Mawson JT, Zamudio S, Moore LG (1998) Catecholamine response during 12 days of high altitude exposure (4,300 m) in women. J Appl Physiol 84(4):1151–1157

Milledge JS, Cotes PM (1985) Serum erythropoietin in humans at high altitude and its relation to plasma renin. J Appl Physiol 59(2):360–364. https://doi.org/10.1152/jappl.1985.59.2.360

Monge C (1942) Life in the Andes and chronic mountain sickness. Science 95(2456):79–84. https://doi.org/10.1126/science.95.2456.79

Moraga FA, Flores A, Serra J, Esnaola C, Barriento C (2007) Ginkgo biloba decreases acute mountain sickness in people ascending to high altitude at Ollagüe (3696 m) in northern Chile. Wilderness Environ Med 18(4):251–257. https://doi.org/10.1580/06-WEME-OR-062R2.1

Murdoch DR (1996) Acute mountain sickness. J R Soc Med 89(12):728

Naik SR, Pilgaonkar VW, Panda VS (2006) Evaluation of antioxidant activity of Ginkgo biloba phytosomes in rat brain. Phytother Res 20(11):1013–1016. https://doi.org/10.1002/ptr.1976

Packer L, Tritschler HJ, Wessel K (1997) Neuroprotection by the metabolic antioxidant alpha-lipoic acid. Free Radic Biol Med 22(1–2):359–378. https://doi.org/10.1016/s0891-5849(96)00269-9

Palmer BF (2010) Physiology and pathophysiology with ascent to altitude. Am J Med Sci 340(1):69–77. https://doi.org/10.1097/MAJ.0b013e3181d3cdbe

Panossian A, Wikman G, Sarris J (2010) Rosenroot (Rhodiola rosea): traditional use, chemical composition, pharmacology and clinical efficacy. Phytomedicine 17(7):481–493. https://doi.org/10.1016/j.phymed.2010.02.002

Paralikar SJ, Paralikar JH (2010) High-altitude medicine. Indian J Occup Environ Med 14:6–12. https://doi.org/10.4103/0019-5278.64608. https://www.ijoem.com/text.asp?2010/14/1/6/64608

van Patot MC, Keyes LE, Leadbetter G 3rd, Hackett PH (2009) Ginkgo biloba for prevention of acute mountain sickness: does it work? High Alt Med Biol 10(1):33–43. https://doi.org/10.1089/ham.2008.1085

Pérez-Pinzón MA, Xu GP, Mumford PL, Dietrich WD, Rosenthal M, Sick TJ (1997) Rapid ischemic preconditioning protects rats from cerebral anoxia/ischemia. Adv Exp Med Biol 428:155–161. https://doi.org/10.1007/978-1-4615-5399-1_22

Pfeiffer JM, Askew EW, Roberts DE, Wood SM, Benson JE, Johnson SC, Freedman MS (1999) Effect of antioxidant supplementation on urine and blood markers of oxidative stress during extended moderate-altitude training. Wilderness Environ Med 10(2):66–74. https://doi.org/10.1580/1080-6032(1999)010[0066:eoasou]2.3.co;2

Pialoux V, Mounier R, Brown AD, Steinback CD, Rawling JM, Poulin MJ (2009) Relationship between oxidative stress and HIF-1 alpha mRNA during sustained hypoxia in humans. Free Radic Biol Med 46(2):321–326. https://doi.org/10.1016/j.freeradbiomed.2008.10.047

Pierre S, Jamme I, Droy-Lefaix M-T, Nouvelot A, Maixent J-M (1999) Ginkgo biloba extract (EGb-761) protects Na,K-ATPase activity during cerebral ischemia in mice. Neuroreport 10:47–51

Ren X, Zhang Q, Wang H, Man C, Hong H, Chen L, Li T, Ye P (2015) Effect of intravenous iron supplementation on acute mountain sickness: a preliminary randomized controlled study. Med Sci Monit 15(21):2050–2057. https://doi.org/10.12659/MSM.891182

Roach RC, Hackett PH (2001) Frontiers of hypoxia research: acute mountain sickness. J Exp Biol 204(Pt 18):3161–3170

Roncin JP, Schwartz F, D'Arbigny P (1996) EGb 761 in control of acute mountain sickness and vascular reactivity to cold exposure. Aviat Space Environ Med 67(5):445–452

Salazar H, Swanson J, Mozo K, White AC Jr, Cabada MM (2012) Acute mountain sickness impact among travelers to Cusco. Peru. J Travel Med 19(4):220–225. https://doi.org/10.1111/j.1708-8305.2012.00606.x

Scherrer U, Rexhaj E, Jayet PY, Allemann Y, Sartori C (2010) New insights in the pathogenesis of high-altitude pulmonary edema. Prog Cardiovasc Dis 52(6):485–492. https://doi.org/10.1016/j.pcad.2010.02.004

Schilling L, Wahl M (1999) Mediators of cerebral edema. Adv Exp Med Biol 474:123–141. https://doi.org/10.1007/978-1-4615-4711-2_11

Schmidt MC, Askew EW, Roberts DE, Prior RL, Ensign WY Jr, Hesslink RE Jr (2002) Oxidative stress in humans training in a cold, moderate altitude environment and their response to a phytochemical antioxidant supplement. Wilderness Environ Med 13:94–105

Seupaul RA, Welch JL, Malka ST, Emmett TW (2012) Pharmacologic prophylaxis for acute mountain sickness: a systematic shortcut review. Ann Emerg Med 59:307–317

Stream JO, Grissom CK (2008) Update on high-altitude pulmonary edema: pathogenesis, prevention, and treatment. Wilderness Environ Med 19:293–303

Talbot NP, Smith TG, Privat C, Nickol AH, Rivera-Ch M, León-Velarde F, Dorrington KL, Robbins PA (2011) Intravenous iron supplementation may protect against acute mountain sickness: a randomized, double-blinded, placebo-controlled trial. High Alt Med Biol 12(3):265–269

Tsai TY, Wang SH, Lee YK, Su YC (2018) Ginkgo biloba extract for prevention of acute mountain sickness: a systematic review and meta-analysis of randomised controlled trials. BMJ Open 8(8):e022005. https://doi.org/10.1136/bmjopen-2018-022005

Van Mil AH, Spilt A, Van Buchem MA, Bollen EL, Teppema L, Westendorp RG, Blauw GJ (2002) Nitric oxide mediates hypoxia-induced cerebral vasodilation in humans. J Appl Physiol 92(3):962–966. https://doi.org/10.1152/japplphysiol.00616.2001

Vardy J, Vardy J, Judge K (2005) Can knowledge protect against acute mountain sickness? J Public Health (Oxf) 27(4):366–370. https://doi.org/10.1093/pubmed/fdi060

Vasankari TJ, Kujala UM, Rusko H, Sarna S, Ahotupa M (1997) The effect of endurance exercise at moderate altitude on serum lipid peroxidation and antioxidative functions in humans. Eur J Appl Physiol Occup Physiol 75(5):396–399. https://doi.org/10.1007/s004210050178

Zafren K (2014) Prevention of high altitude illness. Travel Med Infect Dis 12(1):29–39

Antioxidant Therapy for High Altitude Sickness and Nano-Medicine

13

Pallavi Mudgal and Swati Paliwal

Abstract

High altitude sickness is associated with increased levels of reactive oxygen species (ROS) that leads to oxidative stress in cells. At elevated mountains, hypoxia induces several signaling pathways that contribute to imbalance in cellular homeostasis and oxidative stress. This can further lead to altered redox balance and accumulation of free radicals resulting in numerous pathophysiological disorders including pulmonary asthma, cardiovascular and metabolic disorders. Balanced ratio of pro- and antioxidants is essential for healthy life. Thus, antioxidant supplements are usually given at high altitudes to maintain cellular homeostasis. Use of antioxidant supplements can strengthen the surpassed levels of ROS in respond to oxidative stress and the supply of antioxidants is enhanced by using nanotechnology. The recent advancement in nanotechnology has ensued sustained delivery of drugs and more efficient results in curing various metabolic disorders. Emergence of nanotechnology has provided researchers with enhanced solubility, bioavailability, stability with less toxicity and conventional side effects. This chapter highlights the relation amid high altitudes, increased oxidative stress, and how antioxidant supplements and nanomaterial conjugated antioxidants provide shield to cells and ameliorates the oxidative damage in high altitude related sickness.

Keywords

High altitude sickness · Reactive oxygen species · Antioxidants · Nano-medicine · Cardiovascular · Pulmonary disorders

P. Mudgal · S. Paliwal (✉)
Department of Bioscience and Biotechnology, Banasthali Vidyapith, Banasthali, Rajasthan, India
e-mail: swatipaliwal@banasthali.in

Abbreviations

AMS	Acute mountain sickness
CAT	Catalase
DOX	Doxorubicin
GSSG	Glutathione disulfide
HACE	Hypoxia-induced cerebral edema
HAPE	Hypoxia-induced pulmonary edema
HIF	Hypoxia inducible factor
HRE	Hypoxia responsive elements
HR-NPs	Hypoxia responsive nanoparticles
NOX	Nitrogen oxides
Nrf	Nucleoid related factor
PEG-CNPs	Polyethylene glycol ceria nanoparticles
PGC	Pparg coactivator
PHD	Prolyl hydroxylases family
PUFA	Polyunsaturated fatty acid
ROS/RNS	Reactive oxygen species/reactive nitrogen species
SOD	Superoxide dismutase
TBAR	Thiobarbituric acid reactive substances

13.1 Introduction

High range of foothills has always been a good source of captivation and motivation for populaces. The tranquility and quietness of mountains magnetize individuals to explore the environment and trap people to inhabit the areas of even higher spreads of mountains. Investigators have proposed the additional antioxidant supplements for the eminent levels of oxidative stress which could be particularly advantageous in this situation. There is a very précised research done, which suggests that antioxidant supply is beneficial for acute mountain sickness, reducing muscle soreness and improving cell membrane fluidity. The antioxidant supplementation challenge at altitude deceits in defining the perfect amalgamation and concentration of antioxidant nutrients, which controls excess oxidative stress allowing adaptations to hypoxia. Vitamin C and E, β-carotene, selenium are rich antioxidants which have proven beneficiary effects in AMS, muscle soreness, and oxygenation of peripheral tissues. A balanced antioxidants supply can counteract this oxidative stress maintaining ROS levels in the body. The antioxidant supplementation system is enhanced via nanotechnology.

Nanotechnology comprehends advanced technology combined with biological processes comprising manufacturing of new drugs and delivery, reformative medicine, and a defensible environment (Bowman and Hodge 2007; Roco 2003; Sahoo et al. 2007). However, the originality behind the term is, an emerging family of various technologies involving a broad range of nanoscience and nanotechnology, enables the manipulation of matter at atomic levels (Bowman and Hodge 2007; Ramsden 2016). The National Nanotechnology Initiative (NNI) marked the Global emergence of nanotechnology in January 2000 (Roco 2003). This technology reassures re-construction of the world made by human itself by an expansion of uprising products from machineries to medicines (Sachan and Gupta 2015). Hence aid of nanotechnology in high altitude sickness is the major advancement in the treatment of the related disorders. The following chapter elaborates about the benefits of antioxidant therapy and nano-medicine in acute mountain sickness.

13.2   From Physiological to Molecular: Change of Perspectives

Reviewing the existence of different molecular events, it is clearly very evident that all these patho-physiologies might be interconnected to each other. Individuals with acute high altitude sicknesses express the presence of fluids rushed into the extracellular spaces, in brain or in lungs. There lies molecular dynamics in the existence of hypobaric hypoxia and related disorders. These molecular factors have an imperative part in regulating the production of ROS and also they have a key role in release of antioxidants to suppress the possessions of generated reactive oxygen species.

13.3   Stabilization of HIF in Signaling of ROS During Hypoxic Conditions

HIF-1 substitutes a heterodimer, HIF-1α and HIF-1β oxygen and non-oxygenic regulated. The oxygen dependent HIF complex has prolyl hydroxylases family (PHDs) like PHD1, PHD2, and PHD3. In aerobic condition these PHDs degrade α subunit of HIF, whereas in oxygen deficient condition it results in dimerization and stabilization of HIF. At higher altitudes due to lessened PHD2 activity HIF-1α increases which represents HIFs as the major player of the responses to cells due to limited O_2 supply. Though all the forms of PHD play an essential role in HIF regulation, but it is found that suppression of PHD2 increases the level of HIF where HIF-2α level can be enhanced by the suppression of PHD3 (Appelhoff et al. 2004). To stabilize HIF-1α shushing of PHD2 with siRNAs is necessary in case of normoxic human cells, while no effect on the stability of HIF-1α was observed in epistating the role of PHD1, 3 in both normal and hypoxic conditions (Berra et al. 2003). Hence marking PHD2 as potential oxygen sensor. As already discussed mitochondria are the major site for ROS production which further increases during the hypoxic conditions leading to the redox changes resulting in participation in other transcriptional responses. For the binding activity of HIF-1α DNA, mitochondrial ROS is required (Chandel et al. 1998), specifically complex III stabilizes both the subunits of HIF linking ROS and HIF stabilization.

13.4 Association of PGC-1 α and Sirtuins with HIF-1α in Signaling of ROS

Hypoxia causes the self-destruction of mitochondrial cells via the process of autophagy, reducing ROS and providing a sufficient amount of oxygen to the remaining mitochondrial cells. In muscle cells there is a key relation between the activity of HIF-1α and PGC-1α. It is observed that increase in PGC-1α levels rises up the mitochondrial biogenesis, leading to increased oxygen consumption and HIF-1α stabilization. Sirtuins (SIRT) are dependent on ratio of NAD to NADH, where increased NAD levels activate sirtuins and elevated NADH levels overpower SIRT activity. SIRT1 downregulates the activity of HIF-1α by its deacetylation during hypoxia. Another classes of Sirtuin (SIRT1, SIRT6) downregulates HIF-1 alpha mediated transcription by chromatin binding on HRE (Zhong et al. 2010). Similarly, overexpression of SIRT3 downregulates HIF stabilization and ROS production. Kinases, Sirtuins, and PGC-1alpha are some factors that are essentially regulated by the ROS generated by physical exercises for mitochondrial biogenesis. The modulation occurs by changes in the redox state of the body. PGC-1α reduces ROS production either by activation of antioxidant system or by elevating the number of mitochondria.

13.5 Nuclear Erythroid-Related Factor 2

Nuclear erythroid-related factor 2 is a transcriptional factor which legalizes antioxidant, anti-inflammatory, and observes the redox homeostasis. KEAP1 (Kelch-like erythroid cell-derived protein1) is released via Nrf2 during stress conditions (Oh and Jun 2017; Huang et al. 2002). Once released it is transported to the nucleus, activating genes conferring resistance to various oxidative stress related neurodegenerative molecules. It is well established that fluctuations in cellular oxygen due to hypoxia generated oxidative stress affects HIF-1α and Nrf2. Adenocarcinoma cell line (A549) of lungs has elevated NOX1 levels which are required for amplified ROS accumulation during sporadic hypoxia inducing HIF-1α and Nrf2. Inhibition of endogenous NOX1 hinders the expression of Nrf2 and Trx1, whereas its overexpression causes an upsurge in Nrf2 and Trx1. The twin upregulation of these factors upsurges HIF-1α signaling. Hence, Trx1tends to appear as a link between Nrf2 and HIF-1α. Nrf2 is a transcriptional factor which controls the genes of antioxidant and detoxification system (Malec et al. 2010). There has been a reduction in mitochondrial components among the cells surviving H_2O_2 treatment which was prohibited by overexpression of Nrf2 preventing mitochondrial-related morphological changes.

13.6 Oxidative Stress Markers at High Altitude

Increased oxidative stress damages the cellular molecules imposing toxicological implications. To mark the existence of ROS produced due to oxidative stress different indicators are studied. These are called "biomarkers," measuring the normal and pathogenic processes (Biomarkers Definitions Working Group 2001). Injuries to aerobic cell due to ROS production majorly affect the membrane oxidation called as lipid peroxidation (Debevec et al. 2017). This membrane damage causes elevation in exhalation of pentane gas. During hypoxia, polyunsaturated fatty acids (PUFAs) in membranes get highly affected, which alters the normal functions of cells (Magalhães et al. 2005). Another cellular damage is the changes in fluidity of red cell membrane due to vitamin E/depletion in the same (Simon-Schnass 1994).

Training at high altitude influences the antioxidant defense system (SOD and CAT) of RBCs (Güzel et al. 2000). Along with this the increased thiobarbituric acid reactive substances (TBARS) in blood plasma due to lipid peroxidation are linked with the injury in muscle cell membranes (Wozniak et al. 2001; Ramazan et al. 2000; Bernabucci et al. 2002; Vani et al. 2010). Intermittent hypoxia activates HIF with augmented levels of endothelin causing vasoconstriction, inflammatory cytokines, and irregular lipid metabolism (Friedman et al. 2014; Gangwar et al. 2020). There is no inconsistency about the reduction of glutathione reductase and increased glutathione oxidase during hypoxic condition. A potent indicator for oxidative stress is glutathione sulfide (GSSG) as it increases at high altitude (Magalhaes et al. 2004). Also reduced glutathione peroxidase, cytochrome c oxidase, and superoxide dismutase in the lungs appear to be wise indicators for oxidative stress (Lemoine et al. 2018).

13.7 Physiological Consequences of Oxide Generated Stress

Formation of free radical is an important characteristic involved in the complicated patho-physiology of high altitude ailment (Purkayastha et al. 1999). It is reported that exercise before few hours of being exposed to high range marks the increase in sincerity of altitude sickness (Roach et al. 2000). Moller et al. (2001a, b) deduced that workout at high altitude proliferates the breaking of DNA strands which results in hypoxia and depletion in an antioxidant system capability to endure the affront of oxidative stress. Apart from mountain sickness the increased oxidative stress can also be responsible for impairment in oxygen consumption, functioning of muscle, and ultimately contributing to chronic diseases. Figure 13.1 explains the physiology behind hypoxic hypobaric.

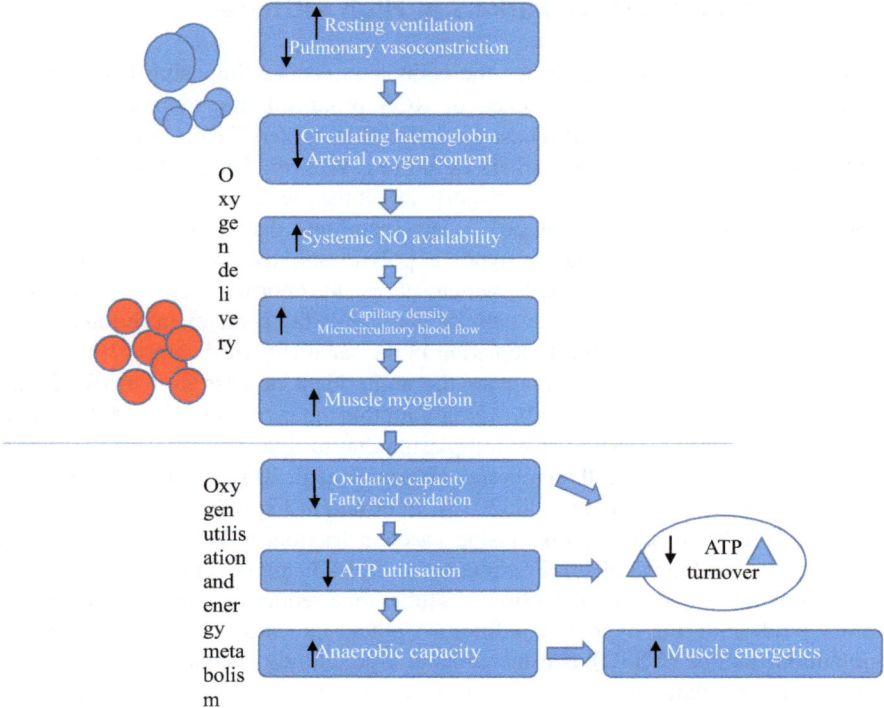

Fig. 13.1 Summarized pathophysiological adaptations to high altitude hypobaric hypoxia

13.8   Antioxidant Therapy for High Altitude Sickness

Higher ROS levels are related with exercise at moderate levels, reducing antioxidant potentiality and increased oxidative stress. The overproduction of ROS, in antioxidant defense systems, damages lipids, proteins, and DNA impairing cell and function of immune system (Fig. 13.2). Fascinatingly, increase in oxidative stress is the consequence of normobaric hypoxia than hypobaric hypoxia. At moderate altitudes, studies have represented inflammation and disease in association with increased oxidative stress. Exogenous antioxidants counterbalance the free radicals, it is rational to theorize that supplementation of antioxidants would be a well-intentioned therapy to battle oxidative stress induced due to high altitude. Though early surveys have shown that supplement of antioxidant modulates the effects on oxidative stress and symptoms of altitude sickness. With the present crucial role of RONS in endurance training, hypoxia and activated antioxidant defenses together, there is no adequate confirmation which recommends that a single dose of antioxidant supplement is sufficient to reduce the oxidative stress generated. Together, there is no adequate confirmation which recommends that single dose of antioxidant supplement is appropriate to attenuate induced stress.

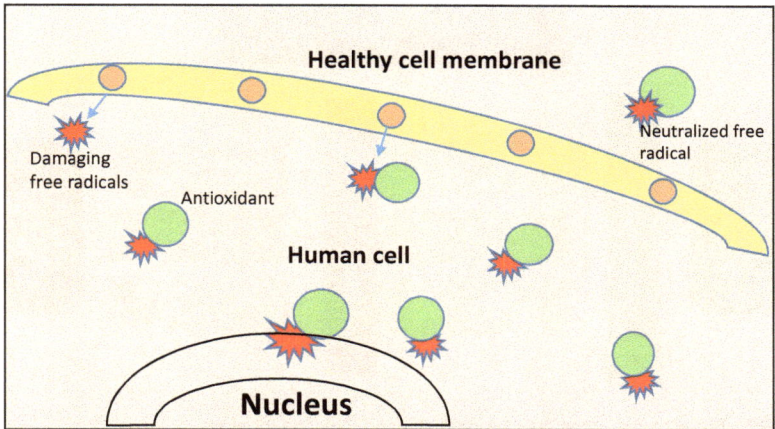

Fig. 13.2 Mechanisms of antioxidant against free radicals

13.9   Effects on Antioxidant System at High Altitude

For normalizing the reactive oxygen/nitrate species effects, an enzymatic and non-enzymatic system has been established by aerobic cells. This system comprises MnSOD, CuSOD which converts superoxide to less empowering hydrogen peroxide species. Along with this glutathione peroxidase and catalase decompose hydrogen species to water. The other non-enzymatic system is complex and consists of non-enzymatic antioxidants. A study reported alternating disclosure to high range at 4000 m, which ensued decrease in protein content activity of mitochondria (Radak et al. 1994). Nakanishi et al. 1995 found that at high altitude the activity of glutathione peroxidase (GPX) decreases in liver signifying the sensitivity of liver to oxidative stress at higher ranges. In other study the action of GPX was compared in the blood serum of highlander's native (Imai et al. 1995). This suggests that GPX activity strongly depends upon the state of thiol system. Glutamyl cycle continuously synthesizes most essential antioxidant, i.e. glutamyl cysteinyl glycine (Fig. 13.3). The level of reduced glutathione decreases as the altitude range increases, while the level of oxidized glutathione increases with increased altitude range (Ilavazhagan et al. 2001; Joanny et al. 2001). All these studies represent that there is a heavy decrease in the capacity of enzymatic systems at higher altitudes. Schmidt et al. (2002) applied an antioxidant combination for the reduction in oxidative stress due to altitude. This combination was found to be very active and reduced the damage caused by increased oxidative stress. Few rats were shortly exposed to an altitude range of 8000 m, later it was found that melatonin levels were increased in their blood serum (Kaur et al. 2002). This melatonin acts as an antioxidant, in which exposure after 4 days reported the decrease in mitochondrial number of pinealocytes which suggested that there is another source apart from pinealocytes which also produce melatonin. There are some pointers which suggest that

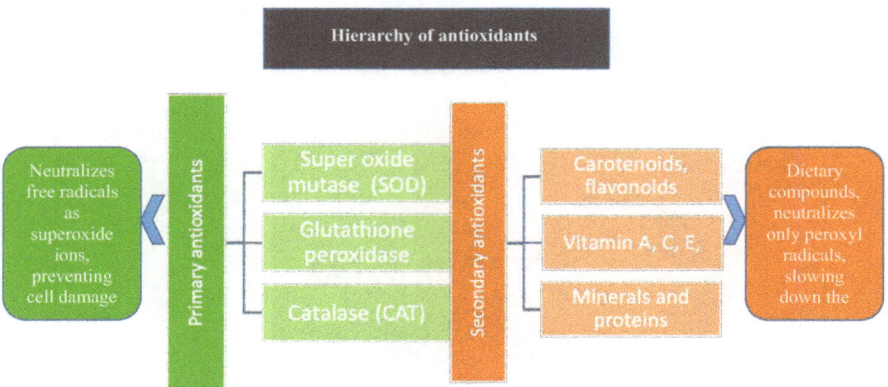

Fig. 13.3 Primary and secondary antioxidants and their functioning

antioxidant supplementation thwarts the oxidative damage due to high altitude to macromolecules.

13.10 Status of Antioxidant Defense System in Body

For the survival of all forms of life specifically aerobic in nature, detoxification of ROS is an essential factor to pay attention to. In order to enhance the deleterious effects produced by oxidant species, the human system is armed with collection of antioxidants (Table 13.1), which are characterized as enzymatic and non-enzymatic, supplied exogenously via food. These supplements actively produce free radical scavengers by contributing electrons to ROS (Kunwar and Priyadarsini 2011; Shinde et al. 2012; Birben et al. 2012).

The antioxidant system works differently in the body: (1) formation of ROS at minimum level, (2) scavenging of reactive species via catalytic molecule, (3) repair and removal of damaged molecules (Sies 1986). This system holds the capacity to develop antioxidants on adaptation to stress environment, in an appropriate concentration. Hence, it is essential for a chain reaction to occur completely in order to stabilize the generated radicals via stearic hindrance. This determines the importance of antioxidant's efficacy. It is witnessed that antioxidant defense system deteriorates at high range, which can be conquered by supplying antioxidants as food add-ons to the body (Poljsak et al. 2013; Halliwell 2011).

13.11 Antioxidant Therapy: Prevention of High Altitude Sickness

Altered endovascular permeability due to excessive ROS generation results in the pathophysiological conditions as AMS, HAPE, and HACE hindered by the supplementation of antioxidants, respectively. These antioxidants could be descent oxygen

Table 13.1 Biological antioxidant defense system

S. No.	Antioxidant system	Key function
1	Vitamin E (α-tocopherol)	Interference with chain reactions mediated via free radicals
2	Vitamin C (ascorbic acid)	Transforms vitamin E free radicals, disturbing lipid peroxidation
3	Glutathione stimulating hormone (GSH)	Reduction in lipid and hydrogen peroxide, oxidation of GSH to GSSG
4	Flavonoids	Potential of donating electrons or hydrogen atoms, inhibition of lipid peroxidation
5	Vitamin A	Quencher of singlet oxygen
6	Uric acid	Singlet oxygen quencher, scavenges radicals
7	Plasma proteins	Prevents formation of HO
8	Superoxide dismutases (SOD)	Dismutate superoxide anions to H_2O_2
9	Glutathione peroxidase (GPx)	Reduction of hydrogen and lipid hydro-peroxides
10	Catalase	Hydrogen peroxide reduction
11	Peroxiredoxin	Alcoholic forms by peroxide reduction
12	Thioredoxin (Trx)	Reduces to H_2O_2 oxygen and water
13	Thioredoxin reductase	Reduction of oxidized Trx
14	Glutathione transferase	Inactivation of secondary metabolites
15	Glutathione reductase	Reduction of glutathione oxidized form
16	Glutamate cysteine ligase	Catalysis production of glutathione
17	Repair enzymes for DNA	Correcting the faults due to oxidative damage

or a combinatory therapy in severe illness. The extract of *Ginkgo biloba* consists of antioxidant properties and showed a protective effect on rats with condition of HACE. It was detected that subjects treated with this extract showed diminished MDA levels along with an increased SOD and GSH concentration (Botao et al. 2013). Patir et al. (2012) explained the part quercetin plays in reduction of hypoxia-induced cerebral edema (HACE) against rats. The development of AMS at an altitude of 5000 m is directly associated with increased serum hydro-peroxides level. Supplementation of antioxidants as vitamins majorly reduces the oxidative stress, enhances total GSH content during high altitude hypoxia (Magalhaes et al. 2004; Araneda et al. 2005). It was analyzed that supplementation of antioxidants tends to be a better alternative to acetazolamide, a pharmacological drug, which is used to inhibit symptoms of AMS protecting hypoxic tissues with no after effects (Gertsch et al. 2004). Hereafter, it can be said that there are few benefits of these non-pharmacological mediators, but with defined dosages only in order to battle these patho-physiologies at high altitude (Table 13.2).

Table 13.2 Some dietary antioxidant compounds, their sources and benefits

S. No.	Antioxidant compound	Benefits of the compound	Sources of compound
1	Vitamin C	Regulates various metabolic diseases, fights acne, improves vision	Raw peppers, parsley, broccoli, cauliflower, fruits rich in citrus, berries, lettuce, sprouts, papaya
2	Vitamin E	Aids in cardiovascular and circulatory system	Sunflower seeds, almonds, vitamin E supplements
3	Coenzyme Q10	Provides cellular energy, helps in maintaining glucose level	Ubiquinol supplements, fish, meat
4	Glutathione	Works with the respiratory system, vision, and immune system	Spinach, potatoes, asparagus, avocado, squash, cauliflower, walnuts, garlic
5	Alpha lipoic acid	Fights against diabetes	Supplements, broccoli, green vegetables
6	Selenium		Brazil nuts, selenium supplements, shrimp, calves liver, salmon
7	Beta carotene	Prevents pain in joints, arthritis condition, and protection against radiations	Chlorella, spirulina, cooked carrots, pumpkin, sweet potatoes, cantaloupe
8	Zeaxanthine		Cooked green peas, romaine, Brussels, corn, broccoli, sprouts
9	Lycopene	Prevents condition of cataract and macular degeneration	Cooked and raw tomatoes, watermelon, guava, pink grapefruit
10	Astaxanthin	Prevents exercise induced asthma and prostate cancer	Astaxanthin, salmon, shrimp
11	Flavonoids	Aids in neurological disorders, provides stamina	Dry beans, fruits, vegetables, green tea, herbs, and spices
12	Quercetin	Helps in asthma and related allergies	Onions, chives, leeks, scallions, garlic
13	Hesperidin	Works as an anti-inflammatory and anti-histamine agent	Apricots, buckwheat, cherries, prunes, rose hips, citrus fruits
14	Curcumin	Most potential antioxidant, antibacterial, and anti-inflammatory properties	Turmeric, curcumin supplements
15	*Ginkgo biloba*	Helps in circulation and proper functioning of brain	Supplements
16	Anthocyanins	Vision and brain functioning	Acai, goji berries, mangosteen
17	Pycnogenol	Relieves joint pain	Supplements
18	Resveratrol	Improves immune system	Muscadine grape seeds, supplements, organic red wine
19	Bilberry	Specifically works against night blindness	Bilberries, supplements
20	Milk thistle	Detoxifies liver and boosts glutathione levels	Supplements

13.12 Nano-Medicine in Acute Mountain Sickness

13.12.1 Beginning of Nanotechnology

Modern science of nanotechnology has been in use for centuries as the nano-materials were utilized to decorate the cathedral windows in medieval times. Not limiting the use of nano-formulations, Chinese used the nano-formulations of gold for introducing red color to ceramic porcelains, acting as an inorganic dye (Pokropivny et al. 2007). In ancient period a process known as Ayurvedic Bhasma was generally used for the preparation of active nanoparticles which involved the melting of metals and then refrigerating them in appropriate media as phyto-herbal juices for definite time. In order to transform these metals into biologically active nanoparticles the above process should be performed numerous times to obtain bhasma (incinerated metals) (Kumar Pal 2015; Sharma and Prajapati 2016). In 1974 Norio Taniguchi coined the term nanotechnology (Allhoff et al. 2010; Krukemeyer et al. 2015), at the University of Tokyo. According to him he described nanotechnology as a resource of yielding particles with accuracy and ultrafine dimensions (Allhoff et al. 2010) (Fig. 13.4). The term nanotechnology is a Greek derived word where "nano" means dwarf. The particle with this technology should have at least one dimension in nanometer (nm) range. Hence, nanotechnology is carved as "technology at nanoscale (1–100 nm)," involving the design and applied materials by regulating their structure and properties (Fakruddin et al. 2012; Ramsden 2016). This field of technology has given promising results in numerous sectors, from health care industry to microelectronics. Specifically in medication, it has shown groundbreaking possibilities in delivering drugs therapies in the areas of research and development (Jena et al. 2017; Roco 2003; Safari and Zarnegar 2014).

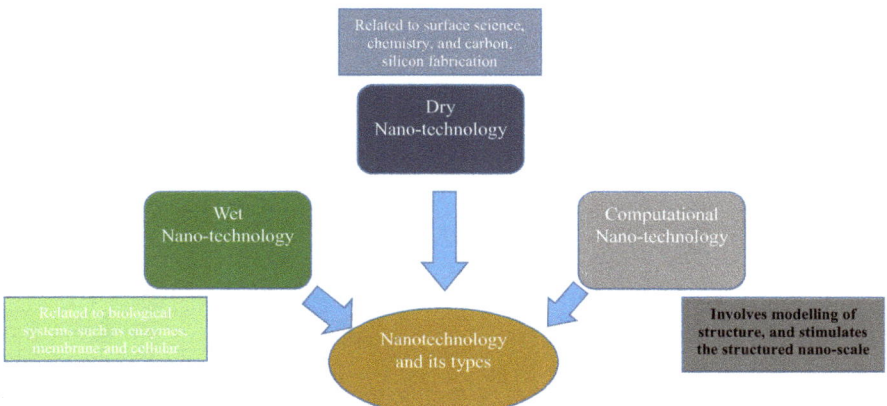

Fig. 13.4 Nanotechnology and its three types

13.12.2 Nano-Formulations

For the application of nanotechnology there has been a consistent development among the formulations of different nano-platforms as nano-biomaterials of a specific surface properties, essential for the interaction of biological compounds and their substantial beneficial effects (Fig. 13.5) (Prasad et al. 2018). Natural and synthetic biomaterials interacting with biological systems have been extensively used in the field of pharmacology as bone grafts, drug transport, tissue engineering. However the advancement in expansion of novel technology has been accomplished by combining the aids of nanotechnology and nano-biomaterials (Lee and Kim 2014). These nano-biomaterials include nanoscaled materials used in the arena of biomedical as delivery of drugs, bio-imaging, tissue engineering, and biosensor (Ali et al. 2013; Sitharaman 2016; Shen 2006; Yang et al. 2011). Lately, the development of nano-formulations for medicinal drugs has fascinated the attention of several researchers, specifically for delivering drugs and to enhance properties of conventional site directed drugs (Jeevanandam et al. 2016). Dendrimers, nanoparticles synthesized of polymers, liposomes, and micelles are some of the common nano-formulations, attaining significance in the pharmaceutical business for improved drug delivery (Singh et al. 2016).

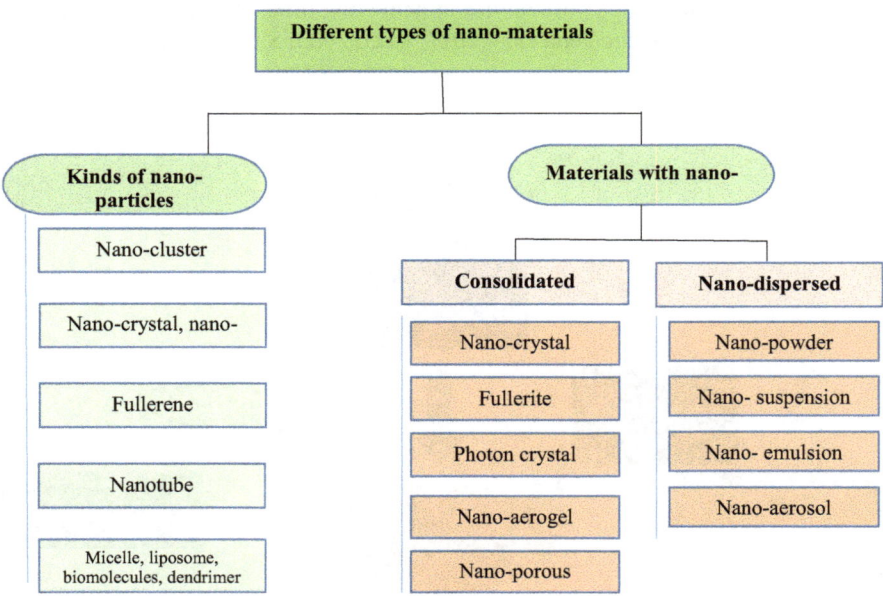

Fig. 13.5 Different kinds of nano-formulations

13.12.3 Hypoxic Nano-Formulations

Low oxygen availability with substantial effects on cells is a condition of hypoxia (Semenza 2015). The affected tissues exist with various conditions when in hypoxic situation (Airley et al. 2000), as disruptive sleep, cerebral disorders, and several cardiac disorders (Bhatia et al. 2017). Hypoxic conditions come with inflammatory responses as rheumatoid arthritis, bowel disease, and ischemic reperfusion injury (Airley et al. 2000). The condition of hypoxia (increased ROS levels) has become a threat to life. It can be fatal for both healthy individuals and individuals with cardiovascular, respiratory and hemolytic diseases (Sun et al. 2016). The application of nanotechnology is to develop nano-formulations to overpower the resistance of delivering conventional drugs and herbal formulations for various pathophysiological conditions including respiratory, cardiac myotrophy, malignancy, and high altitude induced conditions.

13.12.4 Hypobaric Hypoxia and Nano-Formulations

As discussed earlier, hypobaric hypoxia fails to adapt the exposure of high altitudes which in turn is allied with different physiological disorders as pulmonary edema (HAPE) and cardiac hypertrophy (HACE) (Wilkins et al. 2015). For the upgrading of treatment for the mountain illness few investigators have been working in the arena of nanotechnology for the development of nano-therapeutics against high altitude physiologies. Such therapeutic nano-formulations include nano-curcumin and nano-ceria.

13.12.5 Nano-Curcumin for Hypobaric Hypoxia

Though native curcumin has many pharmacological properties, but still there exhibits certain restrictions on its properties of being an effective pharmacological drug. To overcome these restrictions different nano-formulations have been developed, demonstrating permeability, protracted blood circulation, better constancy, and précised discharge of a dose at site to be targeted (Gera et al. 2017). There exist two specifics about both hypoxia and curcumin, i.e. cardiomyocyte hypertrophy induced by hypoxia and curcumin with antioxidant and anti-hypertrophic effects to combat the same. The pharmacological ability of curcumin is limited due to its low bioavailability. Nehra et al. (2015) conducted a study to analyze severity of nano-curcumin against hypertrophy and apoptosis induced by hypoxia and compared it to the naked curcumin targeting the same. Consequences of this research revealed that nano-curcumin expressively countered the hypoxia-induced hypertrophy and apoptosis via downregulating the activation of several factors essential for reducing the production of ROS. This study concluded that nano-curcumin can potentially cure cardiac pathologies induced via hypoxia by restoration of oxidative balance (Nehra et al. 2015). Moreover, improvement in the cardiac damage due to chronic hypobaric hypoxia can be done by using nano-curcumin as compared to nano-curcumin, Nehra et al. (2016b). High altitude acquaintance frequently leads to accretion of fluids in lungs, resulting in high altitude-induced pulmonary edema (HAPE) (Sagi et al. 2014). Thus, the above study underlined the defensive worth of nano-curcumin in lungs at high altitudes (Nehra et al. 2016a).

13.12.6 Nanoceria for Hypobaric Hypoxia

Nanoparticles customized of cerium oxide have a quenching effect against reactive oxygen species (ROS) both in vitro and in vivo as well. Although their efficiency in protecting lungs during oxidative stress due to hypobaric hypoxia was left undiscovered until 2013. So, using microemulsion method spherical nanoparticles (7–10 nm) were produced for lung protection during hypobaric hypoxia. Arya et al. (2014) found that nanoceria decreases ROS content by lowering the amount of cellular calcium. Impaired memory and cognitive dysfunction occur when an individual exposed to high altitude ranges develops reactive nitrogen and oxygen species in the cortex and hippocampus area of brain. Nanoceria coated with polyethylene glycol (PEG-CNPs) were proficiently localized in the brain of rodent resulting in reduced oxidative stress and related damage during. Consequently, demonstrating the promising act of nanoceria as therapeutic agent in metabolic diseases (Arya et al. 2016).

13.12.7 Nano-Formulations Forthcoming for Hypobaric Hypoxia

In patients with various metabolic diseases, hypoxia is commonly witnessed, promoting multiple organ failure frequently (Sun et al. 2016). The developed nano-therapeutics are appreciated in the ailment of patho-physiologies induced by high altitude. Attempts have already been made to stabilize HIF-1α, involved in the reaction to hypoxia virtually (Arachchige et al. 2015). Self-assembled hypoxia responsive nanoparticles (HR-NPs) encapsulating doxorubicin (DOX) were developed by Thambi et al. (2014). HR-NPs can successfully supply DOX into human carcinoma cell line under hypoxia. In vivo biodistribution study established that HR-NPs were gathered specifically at the site of hypoxic tumor cells. This study accentuated the possibility of HR-NPs as nano-carriers for targeted delivery in treatment of hypoxic diseases (Thambi et al. 2014). Some instances of drugs conjugated with nano-formulations are listed in Table 13.3, which tend to regulate and prevent high altitude hypoxic physiologies.

Table 13.3 Some nano-formulations examples that are considered as the valuable candidates for managing the high altitude sickness

S. No.	Nano-formulations	Active nano-molecule	Pharmacological action	References
1	Polyethylene glycol loaded (NO)	Nitric oxide	Targeted anti-inflammatory effect	Cabrales et al. (2010)
2	SOD	(D,L-lactide-co-glycolide) (PLGA)	Improves uptake of neurons and neuroprotective effect by SOD nanoparticles After oxidative stress induced via hydrogen peroxide	Reddy et al. (2008)
3	Nuclear factor κB (PEG-PLGA)	NF-κB decoy	Indigenous delivery of NF-κB into lungs, thereby inhibiting the development of pulmonary edema and hypertension	Kimura et al. (2009)
4	Encapsulated chitosan	Minocycline hydrochloride	Improved neuronal uptake for better neuroprotective and neuro-restorative effects in ischemic injury/stroke	Nagpal et al. (2013)
5	Chitosan-nano-composites	Au nanoparticles	Adaptogenic and antioxidant effects	Koryagin et al. (2013)
6	Nano-selenium	Selenium	Better effect against pulmonary arterial hypertension at high altitudes, reduced levels of lipid peroxidation	Moghaddam et al. (2017)
7	ABI-009 or nab-rapamycin	Rapamycin	Boosted engulfment of rapamycin	Anselmo and Mitragotri (2016)
8	Liprostin	Prostaglandin E-1 (PEG-1)	Improvement in drug dynamics	Bulbake et al. (2017)
9	Poly(glycerol-succinic acid)	Encapsulation with Camptothecin	Improved cellular uptake and retention	Morgan et al. (2006)
10	Nanoparticles composed of polymers	*Magnolia officinalis* isolated Honokiol	Boosted vascular management	Zheng et al. (2010)
11	*Ziziphus mauritiana* extract	Ziziphus mauritiana extract	Immuno-modulatory activity in the extract	Bhatia et al. (2011)
12	*Cuscuta chinensis*	Flavonoids and lignans	Improved solubility	Yen et al. (2008)
13	Curcumin	Curcumin from *Curcuma longa*	Greater solubility and bioavailability	Sahu et al., 2008

(continued)

Table 13.3 (continued)

S. No.	Nano-formulations	Active nano-molecule	Pharmacological action	References
14	Quercetin and polyvinyl alcohol	Quercetin	Antioxidant and free radical scavenger	Wu et al. (2008)
15	Liposome	Breviscapine *Erigeron breviscapus* isolated	Deterrence of cerebral and cardiac diseases	Chakraborty et al. (2016)
16	Nano-precipitated naringenin	Naringenin isolated from citrus fruits	Antioxidant and anti-inflammatory	Bilati et al. (2005)
17	Microsphere composed of chitosan	Genistein isolated from soybean	Utilized in cancer complications.	Si et al. (2010)
18	Berberine formulations	Berberine from roots of *Berberis vulgaris*	Inactivation of cyclooxygenase-2 and DNA topoisomerase II	Chakraborty et al. (2016)
19	Silymarin	Silymarin isolated from *Silybum marianum*	Anti-hepatotoxic effect, reduction of lipid content in blood, anti-diabetic.	Xu et al. (2011)
20	Cryptotanshinone	Cryptotanshinone from roots of *salvia milti*	Anti-inflammatory, cytotoxic, antibacterial, anti-parasitic effect	Hu et al. (2010)

13.13 Conclusion

Increment in ROS production on exposure to high altitude disrupts the efficiency of antioxidant system, resulting in damaging macromolecules oxidatively. Supplementation of antioxidant supplementation has advantageous effects that can diminish the oxidative damage allied with high altitude. Surplus levels of ROS majorly affect cell to cell signaling upsetting overall physiology of individuals exposed to high ranges. Appropriate quantity of antioxidant supplementation is an important way to provide control on excessively produced ROS for one to adapt in hypoxic condition.

Nanoscience is an emerging technology with different applications of targeted delivery which is further utilised for the better treatment of human related disorders. Patho-physiologies related to high altitude are a menace to human physiology under hypoxic environment, and when in contact with UV. A few pharmacological intrusions are being utilized for the management of these physiologies under these harsh environmental conditions. It would be more profitable if nano-based products are utilized in this framework, because nano-formulations improve the pharmacokinetics, conserving value of allopathic and herbal drugs. This chapter presented readers, with the concepts of molecular mechanisms behind prevalence of high altitude sickness, antioxidant therapy for counteracting hypobaric hypoxia and nano-formulations enhancing the effects of antioxidants. This chapter is essential for improved conception of high altitude related patho-physiologies and the other tactics applied for their supervision.

References

Airley RE, Monaghan JE, Stratford IJ (2000) Hypoxia and disease: opportunities for novel diagnostic and therapeutic prodrug strategies. Pharm J 264(7094):66

Ali SH, Almaatoq MM, Mohamed AS (2013) Classifications, surface characterization, and standardization of nanobiomaterials. Int J Eng Technol 2(3):187

Allhoff F, Lin P, Moore D (2010) What is nanotechnology and why does it matter? From science to ethics. Wiley-Blackwell, Chichester

Anselmo AC, Mitragotri S (2016) Nanoparticles in the clinic. Bioeng Transl Med 1(1):10–29

Appelhoff RJ, Tian YM, Raval RR, Turley H, Harris AL, Pugh CW, Ratcliffe PJ, Gleadle JM (2004) Differential function of the prolyl hydroxylases PHD1, PHD2, and PHD3 in the regulation of hypoxia inducible factor. J Biol Chem 279:38458–38465

Arachchige MC, Reshetnyak YK, Andreev OA (2015) Advanced targeted nanomedicine. J Biotechnol 202:88–97

Araneda OF, García C, Lagos N, Quiroga G, Cajigal J, Salazar MP, Behn C (2005) Lung oxidative stress as related to exercise and altitude. Lipid peroxidation evidence in exhaled breath condensate: a possible predictor of acute mountain sickness. Eur J Appl Physiol 95:383–390

Arya A, Sethy NK, Das M, Singh SK, Das A, Ujjain SK, Sharma RK, Sharma M, Bhargava K (2014) Cerium oxide nanoparticles prevent apoptosis in primary cortical culture by stabilizing mitochondrial membrane potential. Free Radic Res 48(7):784–793

Arya A, Gangwar A, Singh SK, Roy M, Das M, Sethy NK, Bhargava K (2016) Cerium oxide nanoparticles promote neurogenesis and abrogate hypoxia-induced memory impairment through AMPK-PKC-CBP signaling cascade. Int J Nanomedicine 11:1159–1173

Bernabucci U, Ronchi B, Lacetera N, Nardone A (2002) Markers of oxidative status in plasma and erythrocytes of transition dairy cows during hot season. J Dairy Sci 85:2173–2179

Berra E, Benizri E, Ginouvès A, Volmat V, Roux D, Pouysségur J (2003) HIF prolyl-hydroxylase 2 is the key oxygen sensor setting low steady-state levels of HIF-1a in normoxia. EMBO J 22:4082–4090

Bhatia A, Shard P, Chopra D, Mishra T (2011) Chitosan nanoparticles as carriers of immuno-restoratory plant extract: synthesis, characterization, and immuno-restoratory efficacy. Int J Drug Deliv 3(2):381–385

Bhatia D, Ardekani MS, Shi Q, Movafagh S (2017) Hypoxia and its emerging therapeutics in neurodegenerative, inflammatory and renal diseases. In: Hypoxia and human diseases. InTech, Rijeka, Croatia

Bilati U, Allemann E, Doelker E (2005) Nanoprecipitation versus emulsion-based techniques for the encapsulation of proteins into biodegradable nanoparticles and process-related stability issues. AAPS PharmSciTech 6(4):E594–E604

Biomarkers Definitions Working Group (2001) Biomarkers and surrogate endpoints: preferred definitions and conceptual framework. Clin Pharmacol Ther 69(3):89–95. https://doi.org/10.1067/mcp.2001.113989

Birben E, Sahiner UM, Sackesen C, Erzurum S, Kalayci O (2012) Oxidative stress and antioxidant defense. World Allergy Organ J 5:9–19

Botao Y, Ma J, Xiao W, Xiang Q, Fan K, Hou J, Wu J, Jing W (2013) Protective effect of ginkgolide B on high altitude cerebral edema of rats. High Alt Med Biol 14:61–64

Bowman DM, Hodge GA (2007) A small matter of regulation: an international review of nanotechnology regulation. Columbia Sci Technol Law Rev 8(1):1–36

Bulbake U, Doppalapudi S, Kommineni N, Khan W (2017) Liposomal formulations in clinical use: an updated review. Pharmaceutics 9(2):12

Cabrales P, Han G, Roche C, Nacharaju P, Friedman AJ, Friedman JM (2010) Sustained release nitric oxide from long-lived circulating nanoparticles. Free Radic Biol Med 49(4):530–538

Chakraborty K, Shivakumar A, Ramachandran S (2016) Nanotechnology in herbal medicines: a review. Int J Herb Med 4(3):21–27

Chandel NS, Maltepe E, Goldwasser E, Mathieu CE, Simon MC, Schumacker PT (1998) Mitochondrial reactive oxygen species trigger hypoxia-induced transcription. Proc Natl Acad Sci 95: 11715–11720

Debevec T, Millet GP, Pialoux V (2017) Hypoxia-induced oxidative stress modulation with physical activity. Front Physiol 8:84

Fakruddin M, Hossain Z, Afroz H (2012) Prospects and applications of nanobiotechnology: a medical perspective. J Nanobiotechnol 1(10):1–8

Friedman JK, Nitta CH, Henderson KM, Codianni SJ, Sanchez L, Ramiro-Diaz JM, Howard TA, Giermakowska W, Kanagy NL, Gonzalez Bosc LV (2014) Intermittent hypoxia-induced increases in reactive oxygen species activate NFATc3 increasing endothelin-1 vasoconstrictor reactivity. Vascul Pharmacol 60(1):17–24

Gangwar A, Paul S, Ahmad Y, Bhargava K (2020) Intermittent hypoxia modulates redox homeostasis, lipid metabolism associated inflammatory processes and redox post-translational modifications: benefits at high altitude. Sci Rep 10:7899

Gera M, Sharma N, Ghosh M, Lee SJ, Min T, Kwon T, Jeong DK (2017) Nanoformulations of curcumin: an emerging paradigm for improved remedial application. Oncotarget 8(39): 66680–66698

Gertsch JH, Basnyat B, Johnson EW, Onopa J, Holck PS (2004) Randomised, double blind, placebo controlled comparison of Ginkgo biloba and acetazolamide for prevention of acute mountain sickness among Himalayan trekkers: the prevention of high altitude illness trial (PHAIT). BMJ 328:797

Güzel NA, Sayan H, Erbas D (2000) Effects of moderate altitude on exhaled nitric oxide, erythrocytes lipid peroxidation and superoxide dismutase levels. Jpn J Physiol 50:187–190

Halliwell B (2011) Free radicals and antioxidants-quo vadis? Trends Pharmacol Sci 32:125–130

Hu L, Xing Q, Meng J, Shang C (2010) Preparation and enhanced oral bioavailability of cryptotanshinone-loaded solid lipid nanoparticles. AAPS PharmSciTech 11(2):582–587

Huang HC, Nguyen T, Pickett CB (2002) Phosphorylation of Nrf2 at Ser40 by protein kinase C regulates antioxidant response element mediated transcription. J Biol Chem 277:42769–42742

Ilavazhagan G, Bansal A, Prasad D, Thomas P, Sharma SK, Kain AK, Kumar D, Selvamurthy W (2001) Effect of vitamin E supplementation on hypoxia-induced oxidative damage in male albino rats. Aviat Space Environ Med 72:899–903

Imai H, Kashiwazaki H, Suzuki T, Kabuto M, Himeno S, Watanabe C, Moji K, Kim SW, Rivera JO, Takemoto T (1995) Selenium levels and glutathione peroxidase activities in blood in an Andean high-altitude population. J Nutr Sci Vitaminol (Tokyo) 41:349–361

ISO (2008). ISO 6709:2008(en) preview. www.iso.org. Accessed 8 June 2016

Jeevanandam J, San Chan Y, Danquah MK (2016) Nanoformulations of drugs: recent developments, impact, and challenges. Biochimie 128(129):99–112

Jena M, Mishra S, Jena S, Mishra SS (2017) Nanotechnology future prospect in recent medicine: a review. Int J Basic Clin Pharmacol 2(4):353–359

Joanny P, Steinberg J, Robach P, Richalet JP, Gortan C, Gardette B, Jammes Y (2001) Operation Everest III: the effect of simulated sever hypobaric hypoxia on lipid peroxidation and antioxidant defense systems in human blood at rest and after maximal exercise. Resuscitation 49:307–314

Kaur C, Srinivasan KN, Singh J, Peng CM, Ling EA (2002) Plasma melatonin, pinealocyte morphology, and surface receptors/antigen expression on macrophages/microglia in the pineal gland following a high-altitude exposure. J Neurosci Res 67:533–543

Kimura S, Egashira K, Chen L, Nakano K, Iwata E, Miyagawa M, Tsujimoto H, Hara K, Morishita R, Sueishi K, Tominaga R (2009) Nanoparticle-mediated delivery of nuclear factor κB decoy into lungs ameliorates monocrotaline-induced pulmonary arterial hypertension. Hypertension 53(5):877–883

Koryagin AS, Mochalova AE, Salomatina EV, Eshkova OY, Smirnova LA (2013) Adaptogenic effects of chitosan-gold nanocomposites under simulated hypoxic conditions. Inorg Mater Appl Res 4(2):127–130

Krukemeyer MG, Krenn V, Huebner F, Wagner W, Resch R (2015) History and possible uses of nanomedicine based on nanoparticles and nanotechnological progress. J Nanomed Nanotechnol 6(6):1–7

Kumar Pal S (2015) The Ayurvedic Bhasma: the ancient science of nanomedicine. Rec Pat Nanomed 5(1):12–18

Kunwar A, Priyadarsini KI (2011) Free radicals, oxidative stress and importance of antioxidants in human health. J Med Allied Sci 1:53–60

Lee H, Kim YH (2014) Nanobiomaterials for pharmaceutical and medical applications. Arch Pharm Res 37(1):1–3

Lemoine AJ, Revollo S, Villalpando G, Valverde I, Gonzales M, Laouafa S, Soliz J, Joseph V (2018) Divergent mitochondrial antioxidant activities and lung alveolar architecture in the lungs of rats and mice at high altitude. Front Physiol 9:311

Magalhaes J, Ascensao A, Viscor G, Soares J, Oliveira J, Marques F, Duarte J (2004) Oxidative stress in humans during and after 4 hours of hypoxia at a simulated altitude of 5500 m. Aviat Space Environ Med 75:16–22

Magalhães J, Ascensão A, Marques F, Soares JM, Ferreira R, Neuparth MJ, Duarte JA (2005) Effect of a high-altitude expedition to a Himalayan peak (Pumori, 7,161 m) on plasma and erythrocyte antioxidant profile. Eur J Appl Physiol 93:726–732

Malec V, Gottschald OR, Li S, Rose F, Seeger W, Hänze J (2010) HIF1 alpha signaling is augmented during intermittent hypoxia by induction of the Nrf2 pathway in NOX1-expressing adenocarcinoma A549 cells. Free Radic Biol Med 48:1626–1635

Moghaddam AZ, Hamzekolaei MM, Khajali F, Hassanpour H (2017) Role of selenium from different sources in prevention of pulmonary arterial hypertension syndrome in broiler chickens. Biol Trace Elem Res 180(1):164–170

Moller P, Loft S, Lundby C, Olsen NV (2001a) Acute hypoxia and hypoxic exercise induce DNA strand breaks and oxidative DNA damage in humans. FASEB J 15:1181–1186

Moller P, Loft S, Lundby C, Olsen NV (2001b) Acute hypoxia and hypoxic exercise induce DNA strand breaks and oxidative damage in humans. FASEB J 15:1181–1186

Morgan MT, Nakanishi Y, Kroll DJ, Griset AP, Carnahan MA, Wathier M, Oberlies NH, Manikumar G, Wani MC, Grinstaff MW (2006) Dendrimer-encapsulated camptothecins: increased solubility, cellular uptake, and cellular retention affords enhanced anticancer activity in vitro. Cancer Res 66(24):11913–11921

Nagpal K, Singh SK, Mishra DN (2013) Formulation, optimization, in vivo pharmacokinetic, behavioral, and biochemical estimations of minocycline loaded chitosan nanoparticles for enhanced brain uptake. Chem Pharm Bull 61(3):258–272

Nakanishi K, Tajima F, Nakamura A, Yagura S, Ookawara T, Yamashita H, Suziki K, Taniguchi N, Ohno H (1995) Antioxidant system in hypobaric-hypoxia. J Physiol 489:869–876

Nehra S, Bhardwaj V, Kalra N, Ganju L, Bansal A, Saxena S, Saraswat D (2015) Nanocurcumin protects cardiomyoblasts H9c2 from hypoxia-induced hypertrophy and apoptosis by improving oxidative balance. J Physiol Biochem 71(2):239–251

Nehra S, Bhardwaj V, Bansal A, Saraswat D (2016a) Nanocurcumin accords protection against acute hypobaric hypoxia induced lung injury in rats. J Physiol Biochem 72(4):763–779

Nehra S, Bhardwaj V, Kar S, Saraswat D (2016b) Chronic hypobaric hypoxia induces right ventricular hypertrophy and apoptosis in rats: therapeutic potential of nanocurcumin in improving adaptation. High Alt Med Biol 17(4):342–352

Oh YS, Jun HS (2017) Effects of glucagon-like peptide-1 on oxidative stress and Nrf2 signaling. Int J Mol Sci 19:E26

Patir H, Sarada SKS, Singh S, Mathew T, Singh B, Bansal A (2012) Quercetin as a prophylactic measure against high altitude cerebral edema. Free Radic Biol Med 53:659–668

Pokropivny V, Lohmus R, Hussainova I, Pokropivny A, Vlassov S (2007) Introduction to nanomaterials and nanotechnology. Tartu University Press, Tartu

Poljsak B, Suput D, Milisav I (2013) Achieving the balance between ROS and antioxidants: when to use the synthetic antioxidants. Oxid Med Cell Longev 2013:1–11

Prasad M, Lambe UP, Brar B, Shah I, Manimegalai J, Ranjan K, Rao R, Kumar S, Mahant S, Khurana SK, Iqbal HM (2018) Nanotherapeutics: an insight into health care and multidimensional applications in medical sector of the modern world. Biomed Pharmacother 97(2018): 1521–1537

Purkayastha SS, Sharma RP, Ilavaahagan G, Siridharan K, Ranganathan S, Selvamurthy W (1999) Effect of vitamin C and E in modulating peripheral vascular response to local cold stimulus in man at high altitude. Jpn J Physiol 49:159–167

Radak Z, Lee K, Choi W, Sunoo S, Kizaki T, OhIshi S, Suzuki K, Taniguchi N, Ohno H, Asano K (1994) Oxidative stress induced by intermittent exposure at a simulated altitude of 4000 m decreases mitochondrial superoxide dismutase content in soleus muscle of rats. Eur J Appl Physiol 69:392–395

Ramazan MS, Ekeroglu SH, Dulger H, Algun E (2000) The effect of dietary treatment on erythrocyte lipid peroxidation, superoxide dismutase, glutathione peroxidase, and serum lipid peroxidation in patients with type 2 diabetes mellitus. Clin Biochem 33:669–666

Ramsden J (2016) Nanotechnology: an introduction, William Andrew

Reddy MK, Wu L, Kou W, Ghorpade A, Labhasetwar V (2008) Superoxide dismutase-loaded PLGA nanoparticles protect cultured human neurons under oxidative stress. Appl Biochem Biotechnol 151(2–3):565–577

Roach RC, Maes D, Sandoval D, Robergs RA, Icenogle M, Hinghofer-Szalky HH, Lium D, Loeppky JA (2000) Exercise exacerbates acute mountain sickness at simulated high altitude. J Appl Physiol 88:581–585

Roco MC (2003) Nanotechnology: convergence with modern biology and medicine. Curr Opin Biotechnol 14(3):337–346

Sachan AK, Gupta A (2015) A review on nanosized herbal drugs. Int J Pharm Sci Res 6(3):961–970

Safari J, Zarnegar Z (2014) Advanced drug delivery systems: nanotechnology of health design: a review. J Saudi Chem Soc 18(2):85–99

Sagi SSK, Mathew T, Patir H (2014) Prophylactic administration of curcumin abates the incidence of hypobaric hypoxia induced pulmonary edema in rats: a molecular approach. J Pulm Respir Med 4:1000164

Sahoo SK, Parveen S, Panda JJ (2007) The present and future of nanotechnology in human health care. Nanomedicine 3(1):20–31

Sahu A, Bora U, Kasoju N, Goswami P (2008) Synthesis of novel biodegradable and self-assembling methoxy poly (ethylene glycol)–palmitate nanocarrier for curcumin delivery to cancer cells. Acta Biomater 4(6):1752–1761

Semenza GL (2015) AJP-cell theme: cellular responses to hypoxia. Am J Physiol Cell Physiol 309(6):C349

Schmidt MC, Askew EW, Roberts DE, Prior RL, Ensign WY Jr, Hesslink RE Jr (2002) Oxidative stress in humans training in a cold, moderate altitude environment and their response to a phytochemical antioxidant supplement. Wilderness Environ Med 13:94–105

Sharma R, Prajapati P (2016) Nanotechnology in medicine: leads from Ayurveda. J Pharm Bioallied Sci 8(1):80–81

Shen JC (2006) Nanobiomaterials. Zhongguo Yi Xue Ke Xue Yuan Xue Bao 28(4):472–474

Shinde A, Ganu J, Naik P (2012) Effect of free radicals & antioxidants on oxidative stress: a review. J Dent Allied Sci 1:63–66

Si HY, Li DP, Wang TM, Zhang HL, Ren FY, Xu ZG, Zhao YY (2010) Improving the antitumor effect of genistein with a biocompatible superparamagnetic drug delivery system. J Nanosci Nanotechnol 10(4):2325–2331

Sies H (1986) Biochemistry of oxidative stress. Angew Chem Int Ed Engl 25:1058–1071

Simon-Schnass I (1994) Risk of oxidative stress during exercise at high altitude. In: Sen CK, Packer L, Hanninen O (eds) Exercise and oxygen toxicity. Elsevier, Amsterdam, pp 191–210

Singh K, Ahmad Z, Shakya P, Ansari VA, Kumar A, Zishan M, Arif M (2016) Nano formulation: a novel approach for nose to brain drug delivery. J Chem Pharm Res 8(2):208–215

Sitharaman B (ed) (2016) Nanobiomaterials handbook. CRC, Hoboken

Sun K, Zhang Y, D'Alessandro A, Nemkov T, Song A, Wu H, Liu H, Adebiyi M, Huang A, Wen YE, Bogdanov MV (2016) Sphingosine-1-phosphate promotes erythrocyte glycolysis and oxygen release for adaptation to high-altitude hypoxia. Nature 7:12086

Thambi T, Deepagan VG, Yoon HY, Han HS, Kim SH, Son S, Jo DG, Ahn CH, Suh YD, Kim K, Kwon IC (2014) Hypoxia-responsive polymeric nanoparticles for tumor-targeted drug delivery. Biomaterials 35(5):1735–1743

Vani R, Reddy CS, Asha Devi S (2010) Oxidative stress in erythrocytes: a study on the effect of antioxidant mixtures during intermittent exposures to high altitude. Int J Biometeorol 54:553–562

Wilkins MR, Ghofrani HA, Weissmann N, Aldashev A, Zhao L (2015) Pathophysiology and treatment of high-altitude pulmonary vascular disease. Circulation 131(6):582–590

Wozniak A, Drewa G, Chesy G, Rakowski A, Rozwodowska M, Olszewska D (2001) Effect of altitude training on the peroxidation and antioxidant enzymes in sportsmen. Med Sci Sports Exerc 33:1109–1113

Wu TH, Yen FL, Lin LT, Tsai TR, Lin CC, Cham TM (2008) Preparation, physicochemical characterization, and antioxidant effects of quercetin nanoparticles. Int J Pharm 346(1):160–168

Xu X, Yu J, Tong S, Zhu Y, Cao X (2011) Formulation of silymarin with high efficacy and prolonged action and the preparation method thereof. US Patent 20110201680. 18 August 2011. Jiangsu University

Yang L, Zhang L, Webster TJ (2011) Nanobiomaterials: state of the art and future trends. Adv Eng Mater 13(6):B197–B217

Yen FL, Wu TH, Lin LT, Cham TM, Lin CC (2008) Nanoparticles formulation of Cuscuta chinensis prevents acetaminophen-induced hepatotoxicity in rats. Food Chem Toxicol 46(5):1771–1777

Zheng X, Kan B, Gou M, Fu S, Zhang J, Men K, Chen L, Luo F, Zhao Y, Zhao X, Wei Y (2010) Preparation of MPEG–PLA nanoparticle for honokiol delivery in vitro. Int J Pharm 386(1):262–267

Zhong L, D'Urso A, Toiber D, Sebastian C, Henry RE, Vadysirisack DD, Guimaraes A, Marinelli B, Wikstrom JD, Nir T, Clish CB, Vaitheesvaran B, Iliopoulos O, Kurland I, Dor Y, Weissleder R, Shirihai OS, Ellisen LW, Espinosa JM, Mostoslavsky R (2010) The histone deacetylase Sirt6 regulates glucose homeostasis via Hif1alpha. Cell 140:280–293